Developments in Primatology Progress and Prospects

Series Editor
Louise Barrett, Lethbridge, AB, Canada

AF166698

This book series melds the facts of organic diversity with the continuity of the evolutionary process. The volumes in this series will exemplify the diversity of theoretical perspectives and methodological approaches currently employed by primatologists and physical anthropologists. Specific coverage includes primate behavior in natural habitats and captive settings; primate ecology and conservation; functional morphology and developmental biology of primates; primate systematics; genetic and phenotypic differences among living primates and paleoprimatology.

Sharon L. Gursky • Jatna Supriatna
Angela Achorn

Editors

Ecotourism and Indonesia's Primates

Editors
Sharon L. Gursky
Department of Anthropology, MS 4352
Texas A&M University
College Station, TX, USA

Jatna Supriatna
Department of Biology
Faculty of Mathematics and Sciences
The University of Indonesia
Depok, Indonesia

Angela Achorn
Department of Comparative Medicine
University of Texas MD Anderson Cancer
Center, Keeling Center for Comparative
Medicine and Research
Bastrop, TX, USA

ISSN 1574-3489   ISSN 1574-3497 (electronic)
Developments in Primatology: Progress and Prospects
ISBN 978-3-031-14918-4   ISBN 978-3-031-14919-1 (eBook)
https://doi.org/10.1007/978-3-031-14919-1

This Springer imprint is published by the registered company Springer Nature Switzerland AG
The registered company address is: Gewerbestrasse 11, 6330 Cham, Switzerland

Preface

The irony of writing, or even just co-editing, a volume on wildlife tourism or any aspect of tourism the past couple of years is not lost on us. At times, we wondered if this volume should be better called "How COVID-19 killed the tourism industry."

When we first envisioned this volume, it was early 2019. Wildlife tourism was booming throughout Indonesia. But then the global pandemic hit. We all expected it might clear up in a month or so - three months at most. How wrong we were. As a result of COVID-19, the original formulation for this volume has drastically changed. Initially there were 18 contributors, but now there are only 9. Many of the contributors had planned to return to the field to collect data but, because of COVID-19 travel restrictions, were unable to do so. Gursky has not been in the field in nearly three years, whereas she rarely missed a field season since 1994. In addition to the inability to actually go to the field, many of the contributors withdrew their contributions because they or their family members got sick, they had no Internet at home, children and other family members in the home made it challenging to work, and a myriad of other challenges resulting from the pandemic.

COVID-19 has not only changed this volume but has changed the tourism industry. One report shows that in in 2021, Bali had only 45 international tourists compared to the millions that that visited annually prior to the pandemic. How COVID-19 has affected our respective tourist sites is something that we are all wondering. Gursky communicated with her field assistants through social media, and they informed her that some of the tarsier tourist groups are no longer at their sleeping site. Some of these sleeping sites have been used for decades. Why, or if, the lack of tourists has caused the tarsiers to move to another sleeping site is a question that will be addressed when we can all return to our respective field sites. Perhaps the lack of tourists feeding the tarsiers crickets made them look elsewhere or perhaps it's just temporary? Time will tell. But the one thing that is certain is that COVID-19 has affected the very animals that the ecotourism industry has been trying to protect—just in very unexpected ways. With the lack of tourism, the local communities are no longer receiving money from tourism that they can use for subsistence. Instead, they may have been forced to rely on the forest and its products. Thus, COVID-19 may have caused additional declines in population densities of many forest-dwelling organisms as well as decreases in the forest itself. Due to the ability of organisms to transmit diseases between different types of organisms, animals may be affected by COVID-19, especially animals related to cats and dogs as well as many types of primates.

All of these questions weigh on our minds as tourism has become tremendously reduced, and local communities are having a hard time surviving, especially with the pace of inflation. Weighing the needs of the people with the needs of animals has always been challenging. With COVID-19, it has become an even greater challenge due to fewer resources and difficulty sharing resources.

College Station, TX, USA   Sharon L. Gursky
Depok, Indonesia   Jatna Supriatna
Bastrop, TX, USA   Angela Achorn

Contents

Chapter 1
Tourism and Indonesia's Primates: An Introduction

Angela Achorn, Sharon L. Gursky, and Jatna Supriatna

Abstract Indonesia is known for its remarkably diverse flora and fauna. One illustration of Indonesia's biodiversity is the impressive number of nonhuman primate species it houses. Recent reports suggest there may be up to 61 primate species in Indonesia, of which 38 are endemic (Perwitasari 2021). More conservative estimates report 48 known species, though this still renders Indonesia the country with the third largest number of primate species worldwide (Estrada et al. 2018). Furthermore, though Brazil and Madagascar are believed to harbor more primate species overall (102 and 100 species, respectively (Estrada et al. 2018)), Indonesia houses all major primate groups, including prosimians, monkeys, lesser apes, and great apes.

This book is a multi-authored volume on primate tourism in Indonesia with the goal of presenting the most up-to-date research on this topic. In this introduction chapter, we clarify different terminology pertaining to tourism and provide an overview of the themes that will be explored throughout this volume – namely, the ecological, economic, educational, and ethical aspects of primate tourism.

1.1 Why Indonesia?

Indonesia is known for its remarkably diverse flora and fauna including high levels of endemism. This megadiversity has resulted from complex biogeographic, geological, climatic, and ecological factors over the last 50 million years (Lohman

A. Achorn (✉)
Department of Comparative Medicine, University of Texas MD Anderson Cancer Center, Keeling Center for Comparative Medicine and Research, Bastrop, TX, USA
e-mail: amachorn@mdanderson.org

S. L. Gursky
Department of Anthropology, MS 4352, Texas A&M University, College Station, TX, USA

J. Supriatna
Department of Biology, Faculty of Mathematics and Sciences, The University of Indonesia, Depok, Indonesia

© The Author(s), under exclusive license to Springer Nature Switzerland AG 2022
S. L. Gursky et al. (eds.), *Ecotourism and Indonesia's Primates*, Developments in Primatology: Progress and Prospects, https://doi.org/10.1007/978-3-031-14919-1_1

et al., 2011). Indonesia presently harbors ~10% of the world's flowering plant species, ~25% of the world's fish species, ~17% of the world's bird species, ~16% of the world's reptiles and amphibians, and ~ 12% of the world's mammals (Rhee et al., 2004; "Indonesia - Country Profile: Biodiversity Facts", 2021). There is a distinct biotic transition near the middle of the archipelago, marked by the Wallace Line, in which flora and fauna of western Indonesia resemble those of mainland Asia, and species of eastern Indonesia resemble those of Australia (Johnson et al. 2019). Indonesia is classified as the home of two biodiversity hotspots: Wallacea and Sundaland. To be considered a biodiversity hotspot, an area must have a large proportion of endemic species, and it must have lost more than 70% of its original natural habitat, often due to anthropogenic factors. There are only 36 of these hotspots worldwide ("Biodiversity Hotspots", 2021).

One illustration of Indonesia's biodiversity is the impressive number of nonhuman primate species it houses. Recent reports suggest there may be up to 61 primate species in Indonesia, of which 38 are endemic (Institut Pertanian Bogor, 2021). More conservative estimates report 48 known species, though this still renders Indonesia the country with the third largest number of primate species worldwide (Estrada et al., 2018). Furthermore, though Brazil and Madagascar are believed to harbor more primate species overall (102 and 100 species, respectively (Estrada et al., 2018)), Indonesia houses all major primate groups, including prosimians (slow lorises and tarsiers), monkeys (macaques, langurs, proboscis monkeys), lesser apes (siamangs, gibbons), and great apes (orangutans). These primates are widely distributed throughout the 16,000 plus island archipelago.

1.2   Why Tourism?

The tremendous diversity of primates in Indonesia, coupled with the conservation issues affecting wildlife in that region, has created a crisis in which many of Indonesia's primates are threatened with extinction. In fact, ~83% of Indonesia's primate species are classified as "threatened" (i.e., listed as Vulnerable, Endangered, or Critically Endangered on the IUCN Red List), and 94% of species exhibit population declines (Estrada et al., 2018). One of the biggest impediments faced by Indonesian conservationists is insufficient financial resources to protect endemic biodiversity. If properly managed, tourism could be used to finance conservation activities. This may be especially valuable in Indonesia where tourism is a priority sector of economic development (Antara & Sri Sumarniasih, 2017). Furthermore, Indonesia's astounding biodiversity, including high levels of species richness and endemism, provides countless opportunities for tourists to appreciate natural phenomena.

The idea that all tourism occurring outdoors or involving nature constitutes "ecotourism" is problematic since that term describes a specific approach inspired by sustainable development initiatives (Stronza et al., 2019). Ecotourism is distinguished from other forms of tourism (e.g., wildlife tourism, nature-based tourism, outdoor recreation) by its conservation, education, and development goals. It involves visitors paying to explore fragile, pristine, and relatively undisturbed natural areas. In order to preserve these sites for future generations to experience, outings are intended to be low impact and small scale compared to standard commercial mass tourism. Ecotourism offers ways to educate visitors, provide funds for ecological conservation, directly benefit the economic development and political empowerment of local communities, and foster respect for different cultures and for human rights (Stronza et al., 2019).

Currently, conservation efforts worldwide rely increasingly on ecotourism for financial and political support. National parks agencies worldwide receive as much as 84% of their funding from ecotourism. Ecotourism has become one of the fastest growing sectors of the tourism industry, increasing annually by 10–15% worldwide. Since ecotourism funds a sizable portion of conservation activities, it is important to examine how it affects the habitats, species, and communities it intends to support. The burgeoning use of "ecotourism" throughout Indonesia has prompted this volume in which authors discuss the successes and pitfalls of tourism at sites harboring nonhuman primates. As will be discussed more extensively, all forms of tourism, including but not limited to ecotourism, can have positive, neutral, or negative effects on wildlife and local communities. These effects all merit consideration.

1.3 This Volume

This book is a multiauthored volume on primate tourism in Indonesia with the goal of presenting the most up-to-date research on this topic. Our approach was inspired by Russon and Wallis (2014) whose eminent volume assessed the effects of tourism on primate conservation around the world. Our volume focuses on Indonesia's primates, in particular, by assessing ecological, economic, educational, and ethical aspects of tourism. We invited experts who study wild animals, including those that naturally occur in forests, villages, and other sites, rather than captive animals housed in zoos, wildlife rescue centers, or roadside "attractions." This volume presents research on prosimians, monkeys, and apes in Indonesia. It also includes a chapter on bird tourism in Indonesia to highlight the value of indigenous knowledge in conservation-focused tourism. Most chapters represent case studies from established field sites where enough data have been collected to assess the long-term effects of tourism.

1.4 Ecological Effects of Tourism

If properly managed, tourism offers opportunities to promote the conservation of natural habitats and species, thereby conferring ecological benefits. However, increased human presence may negatively affect the welfare of animals by causing distress or by altering their "natural" behaviors, diet, and sleep patterns. Furthermore, because organisms are all interconnected, disrupting the lives of one species could affect the overall ecosystem. It is therefore important to evaluate the ecological effects of tourism. The ecological effects of tourism are clearly illustrated in the chapters by Gursky (Chap. 5), Bertrand (Chap. 4), and Nijman (Chap. 9).

1.5 Economic Effects of Tourism

Reports indicate that between 2010 and 2019, the number of tourists visiting Indonesia more than doubled, with approximately 7-million tourists in 2010 and over 16-million in 2019. This tourism generated over $18.4 billion in 2019 alone. Indonesia's impressive biodiversity creates valuable opportunities for tourists to experience natural habitats, which could result in increased employment opportunities and revenue for local communities. However, it has been suggested that perhaps only a fraction of the local community will experience economic benefits of tourism. This is because "local communities" are not undifferentiated entities; the diverse relationships and power dynamics within a community result in differential access to opportunities to participate in, and benefit from, tourism (Stronza, 2008; Gezon, 2013). Furthermore, some tourism researchers—including ones who focus on ecotourism specifically—critique the "neoliberalization of nature" through which nonhuman entities are subjected to market-based systems of management and development (Duffy, 2008). They caution against using contemporary practices rooted in a neoliberal capitalist hegemony to fix issues that capitalism helped create. For these reasons, the economic impacts of tourism should be evaluated to determine who is being affected and what those effects actually entail. The economics of wildlife tourism is discussed in the chapters by Nekaris (Chap. 2), Supriatna et al. (Chap. 6), Tamalene et al. (Chap. 10), and Molyneaux (Chap. 3).

1.6 Ethics of Wildlife Tourism

Primate tourism forces humans and nonhuman primates into closer, more frequent contact, which creates additional opportunities for disease transmission. Parasites and pathogens can hinder an animal's survival and reproduction, which is especially problematic for species at risk of extinction like many of Indonesia's primates (Chapman et al., 2005).

Tourism can also affect human livelihoods. Although it may improve the economic situation of local populations through generating revenue, it could have negative sociocultural impacts. For instance, Indonesia is home to over 500 different ethnic groups, but as people are displaced from their homes due to the development of resorts and homestays, cultures and languages might be altered. Additionally, peak tourist seasons may coincide with religious or national holidays, and the need to work (e.g., guides, park staff, resort/homestay staff, and equipment/gear renters) may interfere with Adat (traditional Indonesian law). It is important to assess the effects of tourism pertaining to the livelihoods of endangered primates and/or human communities. The importance of ethics as pertains to wildlife tourism in Indonesia can be seen in the chapters by Howells et al. (Chap. 8) and Gursky (Chap. 5).

1.7 Education and Tourism

According to Ceballos-Lascurain (1988), an early proponent of ecotourism:

> The main point is that the person that practices ecotourism has the opportunity of immersing him or herself in nature in a way that most people cannot enjoy in their routine, urban existences. This person will eventually acquire a consciousness and knowledge of the natural environment, together with its cultural aspects, that will convert [them] into somebody keenly involved in conservation issues (Ceballos-Lascurain, 1988: 13).

Though not all tourism involving primates constitutes ecotourism, all forms of tourism can make education a priority. This is an important goal, since the combination of tourism, conservation, and education creates opportunities to teach about biodiversity and cultural heritage and to instill values that will inspire future involvement in environmental and social issues (Urias & Russo, 2009; Iakovoglou et al., 2015). The importance of education for wildlife tourism is discussed in the chapters by Hanson et al. (Chap. 7), Supriatna et al. (Chap. 6), Tamalene (Chap. 10), and Howells et al. (Chap. 8).

We hope that this volume helps inspire other researchers and lay people who are engaging in ecotourism, in Indonesia and throughout the world, to be mindful of the ecological, ethical, economic, and educational effects of their activities. We also hope that this volume will inspire additional research on this very important topic.

References

Antara M, Sri Sumarniasih M (2017) Role of tourism in economy of Bali and Indonesia. J. Hosp. Tour. Manag. 5(2). https://doi.org/10.15640/jthm.v5n2a4

Biodiversity Hotspots (2021) Retrieved 17 September, from https://www.conservation.org/priorities/biodiversity-hotspots

Ceballos-Lascurain H (1988) The future of ecotourism. Mexico Journal

Chapman C, Gillespie T, Goldberg T (2005) Primates and the ecology of their infectious diseases: how will anthropogenic change affect host-parasite interactions? Evol Anthropol 14(4):134–144. https://doi.org/10.1002/evan.20068

Duffy R (2008) Neoliberalising nature: global networks and ecotourism development in Madagasgar. J Sustain Tour 16(3):327–344. https://doi.org/10.1080/09669580802154124

Estrada A, Garber P, Mittermeier R, Wich S, Gouveia S, Dobrovolski R et al (2018) Primates in peril: the significance of Brazil, Madagascar, Indonesia and the Democratic Republic of the Congo for global primate conservation. PeerJ 6:e4869. https://doi.org/10.7717/peerj.4869

3

Gezon L (2013) Who wins and who loses? Unpacking the "local people" concept in ecotourism: a longitudinal study of community equity in Ankarana, Madagascar. J Sustain Tour 22(5):821–838. https://doi.org/10.1080/09669582.2013.847942

Iakovoglou V, Zaimes GN, Arraiza Bermudez-Canete MP, Garcia JL, Gimenez MC, Calderon-Guerrero C, Ioras F, Abrudan I (2015) Understanding and enhancing ecotourism opportunities through education. International Journal of Social, Behavioral, Educational, Economic, Business, and Industrial Engineering 9(8):2640–2644

Indonesia - Country Profile: Biodiversity Facts (2021) Retrieved 17 September, from https://www.cbd.int/countries/profile/?country=id#facts

Institute Pertanian Bogor (2021) Prof Rd Roro Dyah Perwitasari: Indonesia Has 38 Endemic Primate Species, Most of Them Are in Sulawesi. Retrieved 17 December 2021, from https://ipb.ac.id/news/index/2021/08/prof-rd-roro-dyah-perwitasari-indonesia-has-38-endemic-primate-species-most-of-them-are-in-sulawesi/c2cf825060509029fbaec2ddab738ef6

Johnson T, Budiarjo A, Halperin J, Catharina C, Rachmansah A, Ridlo M et al (2019) Indonesia tropical Forest and biodiversity analysis (FAA 118 & 119). In: Report for country development cooperation strategy (CDCS): 2020–2025. U.S. Agency for International Development, Indonesia

Lohman D, de Bruyn M, Page T, von Rintelen K, Hall R, Ng P et al (2011) Biogeography of the indo-Australian archipelago. Annu Rev Ecol Evol Syst 42(1):205–226. https://doi.org/10.1146/annurev-ecolsys-102710-145001

Rhee S, Kitchener D, Brown T, Merrill R, Dilts R, Tighe S (2004) Report on Biodiversity and Tropical Forests in Indonesia. Submitted in accordance with Foreign Assistance Act Sections 118/119; February 20, 2004. Prepared for USAID/Indonesia

Russon A, Wallis J (2014) Primate tourism: a tool for conservation? 1st edn. Cambridge University Press, Cambridge, UK

Stronza A (2008) Hosts and hosts: the anthropology of community-based ecotourism in the Peruvian Amazon. NAPA Bulletin 23(1):170–190. https://doi.org/10.1525/napa.2005.23.1.170

Stronza A, Hunt C, Fitzgerald L (2019) Ecotourism for conservation? Annu Rev Environ Resour 44(1):229–253. https://doi.org/10.1146/annurev-environ-101718-033046

Urias D, Russo A (2009) Ecotourism as an educational experience. In: AIEA annual meeting presentation, Atlanta, GA, USA, pp 1–8

Chapter 2
Similar Perceptions of National and International Volunteer Ecotourists Contribute to the Conservation of the Critically Endangered Javan Slow Loris in Java, Indonesia

K. A. I. Nekaris, Ariana V. Weldon, Michela Balestri, and Marco Campera

Abstract Volunteers often provide help with data collection in long-term field studies. Since they usually pay for their experience rather than receive compensation, but also contribute to data collection, they are referred to in the travel literature as "volunteer tourists." Understanding the motivations and the factors affecting the performances of volunteer tourists is pivotal for conservation projects. Here, we aim to understand how local and foreign volunteer tourists (hereafter volunteers) construct meaning of their experiences in joining a primate conservation volunteer program in Indonesia. We investigated the volunteer program at the Little Fireface Project (LFP), a charity working in Indonesia since 2011 to protect the Critically Endangered Javan slow loris (*Nycticebus javanicus*). We analyzed the feedback forms and performances of 74 volunteers (31 Indonesians, 43 foreign) from 2013 to September 2020. Via logistic regressions, we determined that limited differences in perception of the field site and nature were found in the content of the feedback forms between Indonesian and foreign volunteers. Volunteers contributed to 5565 h of data collection, corresponding to half of the total sampling effort. The volunteer program evolved from being 100% foreign volunteers in 2013 to being 100% Indonesian volunteers in 2020. Volunteers wanted a warm and friendly environment and appreciated the fact that we presented a wide range of activities providing

Research Highlights We examined a volunteer tourism program on the conservation and ecology of a globally threatened species, the Critically Endangered Javan slow loris, in West Java, Indonesia. We found that volunteer tourists had similar motivations and expectations regardless of their nationality and contributed positively to the success of the project.

K. A. I. Nekaris (✉) · A. V. Weldon · M. Balestri · M. Campera
Nocturnal Primate Research Group, Oxford Brookes University, Oxford, UK
e-mail: anekaris@brookes.ac.uk

transferrable skills, along with the opportunity for career development. Foreign volunteers complained that the field site is in a village, with loud noise, a human-modified environment, and farmers cutting trees in their own land. Indonesian volunteers appreciated this peculiarity of the project, suggesting that living in a village provided important opportunities for socialization and long-term conservation. We provide evidence that the volunteer program at LFP is successful and significantly contributed to the long-term running of the project and to the in situ conservation of Javan slow loris and other threatened species.

Keywords Volunteer tourism · Motivation factors · Demographics · *Nycticebus javanicus*

2.1 Introduction

Long-term field studies have been lauded as a principal way to aid in species conservation. Presence in an area for several or more years allows us to learn more about a species' ecology, especially when species are long-lived as in the case of primates (Chapman et al. 2017). It is also purported to yield other benefits such as reducing hunting, increasing local capacity and economy, and conservation education leading to positive behavior change resulting in long-term conservation of the target species or ecosystem. In order to collect data over multiple years, however, a team of researchers is needed. Increasingly, long-term field sites turn to volunteers to provide the capacity needed for data collection. Since they usually pay for their experience rather than receive compensation, but also contribute to data collection and restoration or conservation of the environment, they are referred to in the travel literature as "volunteer tourists," or alternatively as cultural tourists, nature tourists, sustainable tourists, or ecotourists (Strzelecka et al., 2017; Wearing 2001). These types of tourism address issues including poverty alleviation, wildlife conservation, restoration of environments, as well as benefits for local individuals, environments, and wildlife (Wearing 2001; Gray and Campbell 2007).

Wanting to contribute to the improvement of human livelihoods and wildlife conservation are principal reasons why volunteers join such projects. Whether or not the projects meet the expectations of those participating has been a focus of volunteer ecotourism studies. For example, motivation to volunteer is important and may include volunteers wanting a new challenge, to learn about a new culture, to travel, or to "give something back" (Campbell and Warner 2016; Polus and Bidder 2016). Other volunteers may have more selfish motivations, such as getting a good line on their c.v. (Galley and Clifton 2004). Motivation may be linked to performance, where some volunteers join a project and commit full-time or more hours, whereas others do not engage and may leave the project early due to disappointment. In the case of working with animals, disappointment at not being able to see or touch the species in question can lead to volunteer disappointment (Cousins et al. 2009), even when knowledge that touching the animal can lead to disease transmission (Russon and Susilo 2014). Broad and Jenkins (2008) found a direct relationship between success of the project and a volunteer's willingness to commit to what they called a long-term commitment (defined as four months). Motivations, however, may differ between residents of a country where a volunteer program takes place with those who travel to that program from abroad (Liu and Leung 2019).

Studies focusing on volunteer tourism in the tropics frequently examine foreigners visiting an "exotic" locality to fulfill sustainable aims, including examining the satisfaction of these foreigners regarding their experiences (Coren and Gray 2011; Liu and Leung 2019). Increasingly, however, volunteer tourists come from the tropical countries themselves and are often young adults in a gap year between degrees or are collecting data for a university degree (Broad and Jenkins 2008; Chen, 2016). Although the aims of the volunteering may be the same, the actual process of acclimatizing to a culture or living away from home may still impact the volunteer experience, including for urban volunteers spending the first time in a rural environment (Brightsmith et al. 2008).

Indonesia is particularly rich in volunteer ecotourism experiences and is a top destination for paying volunteer foreign tourists (Walpole and Goodwin 2000; Lorimer 2009). The lure of charismatic species ranging from orangutans to tigers to Komodo dragons has attracted volunteer tourists to projects run by charities, NGOs and universities, and fully dedicated ecotourism programs such as Operation Wallacea (Galley and Clifton 2004; Wieckardt et al. 2020). Orangutan tourism is one of the best developed in Indonesia, but is fraught with challenges, especially for short-term or day visiting tourists, who may seek to touch or hold orangutans or see them as needing human protection (Russon and Susilo 2014). Oktavia et al. (2020) found that in their orangutan ecotourist project, a way around these complications was through empowering local staff and residents through training and recognition. Ecotourism in Indonesia has been advocated as an essential way to improve rural development overall (Nugroho et al. 2016). Still, most literature on animal ecotourism focusses on motivations, expectations, and feedback from foreign rather than domestic tourists.

The purpose of this chapter is to help to fill this gap by understanding how local and foreign volunteer ecotourists construct meaning of their experiences in joining a primate conservation volunteer program. Since 2013, the Little Fireface Project (LFP) in West Java Indonesia has run a volunteer program whereby volunteers join for a minimum of one month to engage in activities surrounding conservation of the Critically Endangered primate, the Javan slow loris (*Nycticebus javanicus*), which persists in a heavily human-dominated landscape on Indonesia's most populous island, Java (Campera et al., 2021). In this chapter, we review the structure of our volunteer program and examine the demographic over a seven-year period. Using volunteer blogs and feedback forms, we examined whether there were any differences in perceptions between foreign and Indonesian volunteers. Using structural equation modeling, we examined a variety of factors linked to positive or negative perceptions of the study site and nature. We also consider volunteers' scientific contribution to the study of slow lorises. We consider these results in the light of best practice for student volunteer ecotourism.

2.2 Materials and Methods

The Little Fireface Project was established in 2011 to study the ecology of Javan slow lorises, nocturnal primates that are endemic to Java. These animals are threatened by illegal trade and habitat loss and are subject to translocations into habitats not suitable for the species (Nekaris 2016). We also study other species such as Javan palm civets, as well as engage in community education and outreach activities to facilitate their conservation (Nekaris 2016). Our field station and activities are based in the areas surrounding Cipaganti, Garut District, Cisurupan, West Java (S7°6′6–7°7′0& E 107°46′0–107°46′5). Just over 1000 people reside in Cipaganti; they are ethnically Sundanese and predominantly Muslim. The economy comes largely from farming (planting, picking, selling, and processing), although entrepreneurial activities in the form of small food shops, repair shops, mobile phone vendors, etc., also occur. Six schools are within walking distance from or within the village, and villagers estimate that the literacy rate is 90%, with most children going to school until 16. The distance between the edge of the village and the boundary of Gunung Puntang protected forest; forest remains on slopes of the ridges that cannot be cultivated at approximately 1300 m from the village, while the first contiguous forest is about 2000 m away from the village. The land in between reaches up to 1750 m asl and is covered with a mosaic of cultivated fields, abandoned fields and bush patches, bamboo patches, tree plantations, and forest patches. Fields are often bordered by a more or less connected tree canopy (Nekaris 2016). The field station is a large house that we divide into a permanent researcher section and a volunteer section. The volunteers have daily access to a large social room, three private bedrooms with bunkbeds, a kitchen, a bathroom, an equipment room, and outdoor space (front garden) with furniture and games. The staff have an office, a guest reception room, two private bedrooms, and a shower room.

Although the study began in 2011, we have only officially recruited volunteers since 2013. To facilitate volunteer recruitment, the project employed a coordinator. By 2014, this role was divided into a research coordinator (looking after animal data collection and management) and a field station/volunteer coordinator (looking after the daily running of the project and recruiting and hosting volunteers). These coordinators are joined by two to three full-time researchers and five trackers, employed as full-time staff, or working on their PhD degrees. Volunteers were recruited in several ways. We have a volunteer application form and handbook on our website, with an email contact to apply. We recruit volunteers through university programs, through fliers that university staff members pass on to students. Our main recruitment is through call outs in English and Bahasa Indonesia on social media, including Facebook, Twitter, and Instagram, the latter of which yields the most volunteer applications.

From 2013 to 2020, we recruited 110 volunteers (42 Indonesians, 68 foreign). The mean age of volunteers was 24.6 ± SD 5.3 (range = 19.0–48.0) years old. We require a minimum of 1-month volunteering and request that volunteers have some experience related to community conservation. The mean duration of stay for volunteers was 3.5 ± SD 3.1 (range = 1.0–18.0) months. Foreign volunteers (mean: 4.3 months; 95% CI: 3.3–5.2 months) stayed significantly more time than Indonesian (mean: 2.6 months; 95% CI: 1.7–3.5 months) volunteers (Mann–Whitney; U = 3.42; p-value<0.001). We allow volunteers to engage with the more social side of our work (potentially with an education, design, or sustainability background) or with the more ecological side (potentially with ecology, forestry, anthropology, or biology background). All volunteers must do both, so ecological volunteers generally do one day of social volunteering, whereas social volunteers engage in one day of ecology studies and five days of social studies. Generally, volunteers work six days a week, with Sundays as a free day to relax or visit the surrounding areas. Volunteers receive training in health and safety, a history of the project, introduction to the village, and training in the data collection of their area of choice. Students are offered to choose from a list of "mini projects," which are projects that can achieve enough data for a dissertation project of various academic levels (BSc or MSc). PhD students are classified as researchers and generally stay one year or more and are not considered in this study.

Volunteers filled in monthly reports under three categories: *Loris Follows, Education, and Other. Loris Follows* asked volunteers to reflect on their behavioral observation shifts over the previous month, noting any special encounters, memorable moments, struggles, and triumphs. *Education* related to nature club, school activities, community events, and other outreach opportunities. *Other* asked volunteers to add any comments for how their time had been, how they continued with their aims for the month, and their physical/emotional feelings over the past month/week. We examined these reports for patterns and specific quotations relating to volunteer experiences.

Of the 110 volunteers, 74 of them (31 Indonesians, 43 foreign) filled in a completed feedback form at the end of their stay. The feedback form included several topics. We asked why volunteers chose LFP. We asked volunteers to evaluate the information they had been given before arrival and if their expectations met with reality, as well as their overall opinion of the field site, including the living conditions, food, and facilities. We asked their opinion about working with local guides and trackers and their perception of engagement with the local community. We also asked them to give their opinion of conservation and research work conducted by LFP and asked about experiences during behavior observations of lorises. For students undertaking their final projects, we asked how supportive LFP staff were in helping them to reach their research goals. Final questions were about LFP overall, if volunteers had any constructive advice and if there was a staff member who was particularly helpful, so we could commend them for their efforts.

We used binary logistic regression to test the difference in the terms used in the feedback forms by Indonesian and foreign volunteers. To evaluate the factors affecting the volunteers' perceptions of environment at the field site and natural environment at LFP, we tested for mediation effects between variables via structural equation modeling (SEM) via IBM Amos 26 software. In this analysis, we used the variables that emerged from the feedback form (Table 2.1) as both dependent and independent variables (exogenous variables), mediating the variables "environment

Table 2.1 18 categories we identified from volunteer feedback forms regarding their experiences at the Little Fireface Project, West Java, Indonesia, including example quotations

Category	Mentions concepts or terms including	Example
Logistics positive	Positive experiences with the field station/facilities/food and with staff/volunteers	"The LFP station is exactly as described in the guide. Everything was great!"
Logistics negative	Negative comments about the field station/facilities/food and staff/volunteers	"Honestly the only thing that I dislike about it was my phone couldn't receive a good signal if it's in first floor"
Natural environment negative	Negative experiences or things they did not like related to the village, forest, or other surrounding environment	"It's been another bad weather week. A couple of my shifts were cancelled before I even left the house or after spending hours sitting in huts without being able to see any lorises!"
Natural environment positive	Positive experiences or things they liked related to the village, forest, or other surrounding environment	"…the view over the city with Mt. Cikuri in the background and the stars you see on a clear night is just beautiful"
Career	Future career prospects, desired career routes, or volunteering to help with their careers	"My time at LFP was packed with experiences learning skills that I can utilize towards a career in environmental research"
Learn/experience	Learning new skills or having new experiences that are unrelated to professional or personal development	"I am interested to learn about methods, camera trap, data logger, and capture. Hopefully I am able to contribute many things to LFP"
Personal growth and development	Gaining personal wisdom, growth, maturity from the experience	"I had so learned with this travel about so many things, I have the feeling that I have grown, this experience was so rich"
Encouraged by others	Mention of joining LFP by encouragement or recommendation from previous volunteers	"I was recommended by a friend who knew I wanted to learn about conservation, develop my field skills and learn more about lorises"
Travel	Mention of travel (to or around Indonesia), visiting another country	"When I knew I was going to a local congress, I thought I could see lorises while in Asia; LFP immediately came to mind"
Agroforestry	Mention of forest or agroforestry, including projects, surveys, farmers, soil, plantations	"I greatly enjoyed assisting the agroforestry project, in which we planted and provided to farmers over 150 trees"
Conservation	Conservation education, programs, resolving problems, endangered	"I fully believe conservation needs to teach conservation problems from all angles and I think this is being done"
Education	Education programs, activities, teaching, teachers, schools	"Education activities have been amazing!!! The puppet shows were wonderful!!! I was so proud of the kids!"

(continued)

Table 2.1 (continued)

Category	Mentions concepts or terms including	Example
Ecology	Ecology, behavior, biology	"The more time I spend here, the more I know about the importance of lorises in forest ecology, and in human lives"
Biodiversity	Biodiversity, ecosystem	"This gives me confidence in moving on to the role of Javan biodiversity in providing essential ecosystem services"
Degree project	Volunteer's degree project, aim or research goal	"I hope I can finish my project about agroforestry mapping the presence of Loris for my undergraduate university study"
Convenience	Volunteers' motivation for choosing LFP as close to home, easy to get to, not expensive	"In my case, it was because LFP is located not too far away from where I live and from my university"
Favorite/ emotional word slow Loris	Describing slow lorises as volunteer's favorites or with emotional/anthropomorphic language	"I saw Fernando and Alomah– Two of my favorite Loris boys who both looked great"
Market (surveys/ trade)	Mention of visiting wildlife markets, animal or wildlife trade, animal sellers, animals sold	"We saw 7 Loris over 2 markets, and a plethora of as civets, ferret badgers, leopard cats, domestic cats, and turtles"

at the study site" and "natural environment" (endogenous variables). We included covariances in case the exogenous variables were correlated. We used maximum-likelihood estimation and bias-corrected 95% confidence intervals to calculate model parameters. We assessed the goodness of fit of our model by chi-square (χ^2) test, root-mean-square error of approximation (RMSEA), and comparative fit index. We sequentially excluded the least significant variables until the model was stable.

This research was approved by the University Research Ethics Committee at Oxford Brookes University, Oxford, UK (#OBUUREC_1718_VN003). Protocols followed the ethical guidelines proposed by the Association of Social Anthropologists of the United Kingdom and Commonwealth. All research and corresponding activities were approved by the Ministry of Research, Technology, and Higher Education of the Republic of Indonesia (KEMENRISTEKDIKTI) (# 104/SIP/FRP/E5/Dit.KI/IV/2018).

2.3   Results

Of the 110 volunteers, 42 were Indonesian and 68 were foreigners from 18 countries (United Kingdom, Ireland, Spain, France, Italy, the Netherlands, Norway, Sweden, Austria, Germany, Australia, USA, Canada, India, Malaysia, Singapore, Korea, and Japan). Sixty-eight were females and 42 were males. Of 98 students (34 Indonesia,

64 foreign) who designated whether they were doing their volunteering as part of a degree, 38% of volunteers were Indonesian with 62% being foreign. Of those studying for their degree, 29% were Indonesian and 71% were foreign. We found no statistical difference in the general terms used in the feedback forms between Indonesian and foreign volunteers (Table 2.2). The only significant difference was that Indonesian volunteers were more encouraged by others to join as a volunteer at LFP, while foreign volunteers' main reason to volunteer at LFP was to travel.

Overall, both Indonesian and foreign volunteers had a positive impression of the logistics of the field site. They mainly enjoyed the fact that they felt comfortable, like being at home, and appreciated when management had group activities to increase a sense of teamwork. Many foreign volunteers said it was the nicest place they had stayed to do field work, being happy with a bed, indoor toilet, good electricity, and access to Internet. Other foreign volunteers had negative comments regarding the field station environment, with volunteers wanting a greater diversity of food, wanting a shower with hot water, wanting reliable fast Internet, and complaining about village noise levels and waste management systems.

Volunteers had both positive and negative perceptions of the natural environment. They generally appreciated the wildlife, landscape, and the community life, with local people inviting them to community events and being very friendly.

Table 2.2 Terms used by Indonesian ($n = 31$) and foreign ($n = 43$) volunteers in their feedback forms. Values are number (and percentage) of volunteers. Beta (β) coefficients and standard errors (SE) are relative to the binary logistic regression tests to see if the terms used differed between Indonesian and foreign volunteers

	Terms used	Indonesian	Foreign	β (SE)
Perceptions	Logistics negative	6 (19.4)	12 (27.9)	−0.48 (0.57)
	Logistics positive	18 (58.1)	22 (51.2)	0.41 (0.37)
	Natural environment negative	18 (58.1)	17 (39.5)	0.75 (0.48)
	Natural environment positive	11 (35.5)	19 (44.2)	−0.36 (0.49)
Gain	Career	4 (12.9)	6 (14.0)	−0.09 (0.69)
	Learning	23 (74.2)	26 (60.5)	0.63 (0.52)
	Personal growth	9 (29.0)	16 (37.2)	−0.37 (0.51)
Reasons	Convenience	4 (12.9)	1 (2.3)	0.08 (0.95)
	Degree project	9 (29.0)	7 (16.3)	0.74 (0.57)
	Encouraged by others[a]	11 (35.5)	4 (9.3)	1.68 (0.65)
	Slow Loris	13 (41.9)	19 (44.2)	−0.09 (0.48)
	Travel[a]	7 (22.6)	24 (55.8)	−1.47 (0.53)
Topic	Agroforestry	8 (25.8)	20 (46.5)	−0.92 (0.51)
	Biodiversity	2 (6.5)	7 (16.3)	−1.04 (0.84)
	Conservation	12 (38.7)	15 (34.9)	0.17 (0.49)
	Ecology	24 (77.4)	26 (60.5)	0.90 (0.53)
	Education	5 (16.1)	13 (30.2)	−0.81 (0.59)
	Wildlife trade	5 (16.1)	7 (16.3)	−0.27 (0.68)

[a]$P < 0.05$

Negative comments from foreign volunteers about the natural environment were largely based on the fact that the project is not in a national park but in an agroforest, and some students expected to see primary rainforest and large trees. Still, these same students were surprised how much biodiversity thrived in these areas despite the human impact. The main gain Indonesian volunteers identified from the experience was learning new skills and expertise. All volunteers mainly chose LFP because they liked slow lorises and wanted to learn more about them. The main topics mentioned by volunteers were ecology, followed by agroforestry and conservation.

The proportion of Indonesian over foreign volunteers increased through time, and as of 2020, all of the volunteers were Indonesian. This transition was also reflected in an increased proportion of data collected by Indonesian volunteers through time (Fig. 2.1). The total sampling effort on the behavioral observations on Javan slow loris by volunteers between 2013 and September 2020 was 5565 h, corresponding to almost 50% of the total amount of data collected by the project over the same period (11,140 h).

The SEM explaining the factors affecting the volunteers' negative perceptions of study site and nature at LFP was stable after excluding the variables agroforestry, conservation, market survey ($\chi^2 = 53.33$, $p = 0.539$; CFI = 1.000; RMSEA = 0.000) (Fig. 2.2). Volunteers who joined LFP as a travel experience gave more negative feedback to the environment at the field site ($\beta = 0.28 \pm$ SE 0.10, $p = 0.004$). Volunteers who joined LFP to do their degree project gave less negative feedback to the environment and nature ($\beta = -0.28 \pm$ SE 0.13, $p = 0.025$). Volunteers who saw opportunities for their career after joining LFP and who mentioned biodiversity

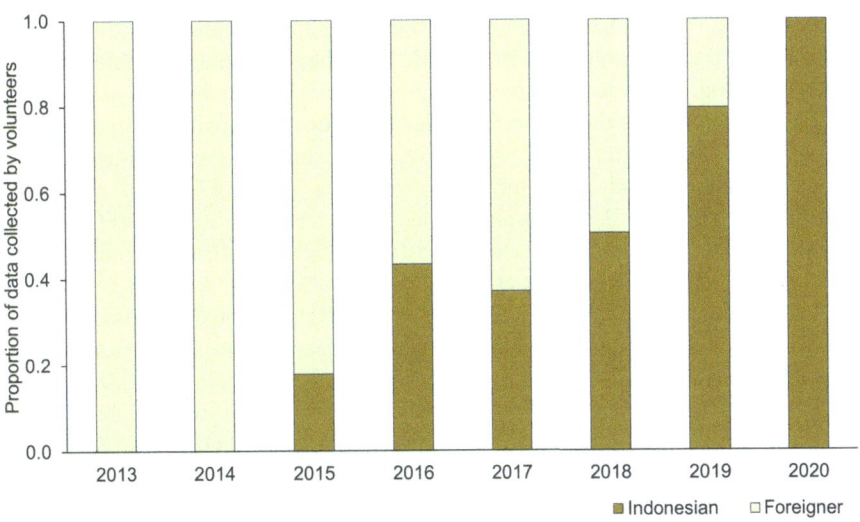

Fig. 2.1 Proportion of slow loris observation data points collected by volunteers (Indonesian vs foreign) on Javan slow loris at the Little Fireface Project between 2013 and 2020

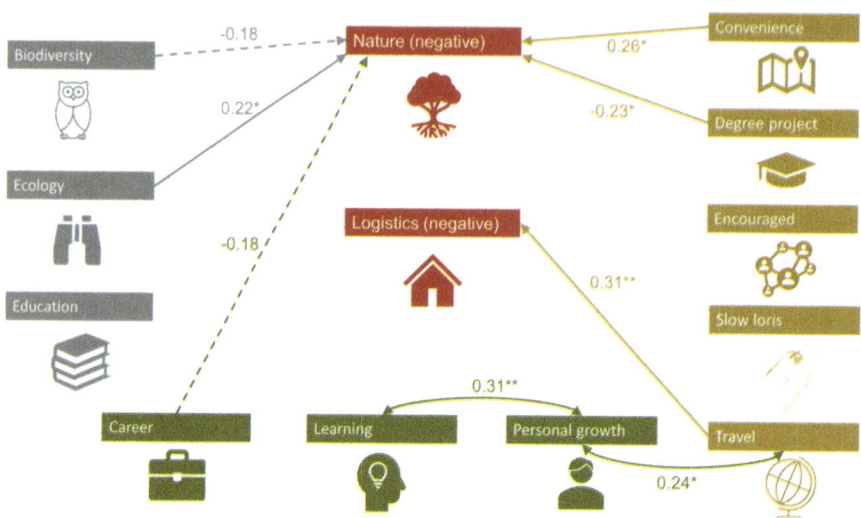

Fig. 2.2 Representation of the structural equation model to understand the factors (exogenous variables) affecting the volunteers' negative perceptions of environment at the field site and natural environment (endogenous variables) at the Little Fireface Project. Values are standardized regression weights. Single arrows indicate that the exogenous variable influence the endogenous variables. Double arrows indicate covariance of exogenous variables. Solid lines indicate significant results ($p < 0.05$); dashed lines indicate tendencies ($0.10 > p > 0.05$). No lines indicate no significant relationships (*$p < 0.05$; **$p < 0.01$)

tended to give less negative feedback to the natural environment (career: $\beta = -0.23 \pm SE\ 0.13$, $p = 0.082$; biodiversity: $\beta = -0.28 \pm SE\ 0.16$, $p = 0.076$). Volunteers who mentioned ecology gave more negative feedback to the natural environment ($\beta = 0.24 \pm SE\ 0.11$, $p = 0.031$). The other exogenous variables did not affect the endogenous variables.

The SEM explaining the factors affecting the volunteers' positive perceptions of the study site and nature at LFP was stable after excluding the variables: degree project, education, and slow loris ($\chi^2 = 37.40$, $p = 0.673$; CFI = 1.000; RMSEA = 0.000) (Fig. 2.3). Volunteers who joined LFP because the field site was in a convenient position or because they were encouraged by others gave more positive feedback to the environment at the field site (convenient: $\beta = 0.48 \pm SE\ 0.18$, $p = 0.009$; encouraged: $\beta = 0.42 \pm SE\ 0.11$, $p < 0.001$). Volunteers who mentioned conservation and ecology also gave more positive feedback to the environment at the field site (conservation: $\beta = 0.28 \pm SE\ 0.10$, $p = 0.004$; ecology: $\beta = 0.26 \pm SE\ 0.10$, $p = 0.010$). Volunteers who saw in LFP an opportunity for personal growth tended to give more positive feedback to the environment at the field site ($\beta = 0.19 \pm SE\ 0.11$, $p = 0.062$). Volunteers who mentioned market surveys gave fewer positive feedback to the environment at the field site ($\beta = -0.48 \pm SE\ 0.15$, $p = 0.001$). Volunteers who saw in LFP an opportunity for personal growth gave more positive feedback to the natural environment ($\beta = 0.46 \pm SE\ 0.11$, $p < 0.001$). Volunteers who mentioned conservation gave more positive feedback to the natural

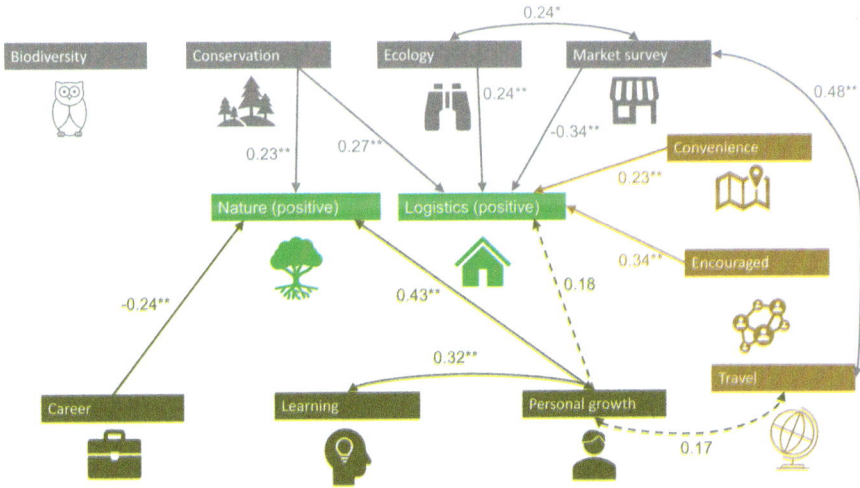

Fig. 2.3 Representation of the structural equation model to understand the factors (exogenous variables) affecting the volunteers' positive perceptions of environment at the field site and natural environment (endogenous variables) at the Little Fireface Project. Values are standardized regression weights. Single arrows indicate that the exogenous variable influence the endogenous variables. Double arrows indicate covariance of exogenous variables. Solid lines indicate significant results ($p < 0.05$); dashed lines indicate tendencies ($0.10 > p > 0.05$). No lines indicate no significant relationships (*$p < 0.05$; **$p < 0.01$)

environment ($\beta = 0.28 \pm$ SE 0.10, $p = 0.004$). Volunteers who saw in LFP an opportunity for their career gave fewer positive feedback to the natural environment ($\beta = -0.36 \pm$ SE 0.14, $p = 0.010$).

2.4 Discussion

Overall, we would suggest that the LFP volunteer program has been a success. At the forefront of this has been providing person power to collect 5565 hours of behavioral data collection on a Critically Endangered species. All volunteers stayed a minimum of one month, providing them with vital training to meet data collection standards (c.f. Broad and Jenkins 2008). At the same time, volunteers noted many positive experiences, and by and large, these did not differ whether the volunteer was local or foreign. These points included the opportunity to be in nature with a rare species, to learn transferable skills, and to grow as a person (Coren and Gray 2011; Alexander 2012). Negative points regarding the environment, including trash in the village, later led to LFP developing trash management programs, and those regarding the need for team building and building relations with villagers also led to increased activities in both areas (Campera et al., in press). These points show that as an organization, LFP could also grow and take inspiration from the observations of the volunteers.

Some observations from the feedback form are of interest when considering the values of young people in the twenty-first century. Feeling in a safe environment has been identified as important for volunteer experiences (Campbell and Warner 2016; Bone and Bone 2018). Despite going to a jungle environment where one might expect to be cut off, many volunteers made comments about the comfort of the accommodation, food preferences, and availability of technology, especially the Internet. It was also evident that many of the volunteers wanted a warm welcome and to be made to feel at home, and, for example, were uncomfortable with areas of the field station being made for staff only. This element of territorialism in volunteers has been identified as a factor that may make volunteers who perceive themselves as unnurtured or unwelcome in a new space, depressed, homesick, or wanting to leave the project (Bone and Bone 2018). Although these areas are considered in our volunteer handbook and volunteers are asked to read this before they come, it could be important for the future to have online meetings with potential volunteers, so they come prepared to know the reality. A debrief at the end of the experience, or a few months after, would also be valuable as we go forward.

Other aspects of value that emerged included the value volunteers felt in learning new technologies and the pride and wonder they described at seeing the slow loris and other animals (Broad and Jenkins 2008; Coren and Gray 2011). Sharing these experiences with local trackers was important for both foreign and Indonesian students, and they valued the experience of staff in their training and as a cultural experience (Chen 2016). At the same time, there was a strong relationship between negative views and ecology. Volunteers were often disappointed, for example, if they could see only the animal's eye shine, or only hear its radio tracking signal from dense shrubbery. At the same time, foreign volunteers felt frustrated at doing tasks such as data entry or habitat mapping as it meant time away from seeing the animals. Cousins et al. (2009) describe this phenomenon not only as frustration but taking away from the exhilaration of other potentially positives experiences. Other studies have reported the disappointment of volunteers who are not prepared for uncomfortable environments, perhaps replete with vermin, cold water baths, and limited technology (Cousins et al. 2009; Campbell and Warner 2016). Similarly, the area where we study is in a human-dominated landscape. Many foreign volunteers complained that they thought they were coming to a forest and were upset when farmers cut trees on their own private land, insisting that LFP "do something." Feelings of hopelessness, anger, or culture shock by volunteers seem to be common across the sector but may also improve when volunteers stay longer (Coren and Gray 2011; Liu and Leung 2019). Anecdotally, but similar to Otoo et al. (2016), we found that three months was a critical period for volunteers to get over these feelings; those that did stayed up to a year or more. Again, materials provided to volunteers about these expectations may help subvert these negative comments.

Motivation to come to the project differed between Indonesian and foreign students. As has been seen in other studies, foreign students were more likely to come for the travel experience (Broad and Jenkins 2008; Alexander 2012). Indonesian students were more motivated to come either with their friends or batchmates or to come after their friend had already done so (Ghose and Kassam 2014). Although some students had a negative impression of the anthropogenic landscape, this changed to a positive one when students were motivated to study for a degree and had a specific topic to focus on, and also for students who wanted to work with the target species (Broad and Jenkins 2008). Even though we offered research topics to non-degree students, they usually decided they wanted to learn "a bit of everything" and then could lose focus or feel frustrated and were more likely to want to be offered entertainment options (Campbell and Warner 2016). They also were likely to stay for a shorter period and fell more into the definition of vacationing ecotourist than a volunteer who helped with species conservation (Strzelecka et al., 2019). For example, they might not understand the reason why we would follow an animal that goes out of view and how those data are still valuable. Fortunately, our project offers many topics for investigation, and many students who joined wanting to do the "ecological" or the "social" side switched focus after actually being in the field and admitted they were surprised in themselves as to what they liked and did not like. Having a list of topics also means that students choose something that is useful for the project.

The aim of our project was always to involve both local and international volunteers. In the first six years of the project, volunteer fees contributed 100% toward the renting and running of the field station and toward the salary of the housekeeper and cook. After 2016, LFP gained sponsorship from two organizations annually that allowed these costs to be paid. This allowed us to provide local rates for Indonesian students. Even after the offer of a competitive local price, we needed to become established in order to attract or become known to local students. In 2018, much more strict permit laws were issued meaning that it became more and more difficult to allow international volunteers to come to LFP. At the same time, the increase in Indonesian volunteers led to a much greater word of mouth transfer of information, making volunteering at LFP a more popular choice. During the 2020 COVID-19 pandemic, when no foreign volunteers could enter the country, and Indonesian volunteers could only stay in their own province, their volunteering became more valuable than ever. Despite several negative effects of the pandemic on international volunteer tourism (Zahawi et al. 2020), we could see this as a positive. We noted also that the confidence in Indonesian volunteers grew and having a larger cohort seemed to foster this positivity (Liu and Leung 2019).

As a small project, we show here that the benefit of volunteers to LFP has been considerable. At the same time, by providing volunteers the opportunity to join the project, they learn invaluable field skills, have access to study globally threatened species, and are able to learn many skills regarding community conservation. This focus on a science with a tangible output of species conservation seems a key strength to improve overall volunteer satisfaction.

Acknowledgments We thank Indonesia RISTEK and the regional Perhutani and Balai Konservasi Sumber Daya Alam for authorizing the study. We thank our field team Daniel Bergin, Helen Birot, Francis Cabana, Rachmatt Cibabuddthea, Katey Hedger, Abdullah Langgeng, Yiyi Nazmi, Adin Nunur, Robert O'Hagan, Kathleen Reinhardt, E. Johanna Rode, Dendi Rustandi, Marie Sigaud, Denise Spaan, and Aconk Zalaeny, as well as the many volunteers who helped Little Fireface Project over the years. This research was funded by Amersfoort Zoo, Augsburg Zoo, Brevard Zoo, Cleveland Zoo and Zoo Society, Columbus Zoo and Aquarium, Cotswolds Wildlife Park, Disney Worldwide Conservation Fund, Global Challenges Fund Initiative—Oxford Brookes University,

Henry Doorly Zoo, International Primate Protection League, Little Fireface Project, Mohamed bin al Zayed Species Conservation Fund (152511813, 182519928), Margot Marsh Biodiversity Fund, Memphis Zoo, Moody Gardens Zoo, National Geographic (GEFNE101-13), Paradise Wildlife Park, People's Trust for Endangered Species, Phoenix Zoo, Primate Action Fund, Shaldon Wildlife Trust, Sophie Danforth Conservation Biology Fund, Zoologische Gesellschaft für Arten- und Populationsschutz.

Conflict of Interests   The authors have no conflicts of interest to declare.

References

Alexander Z (2012) The impact of a volunteer tourism experience, in South Africa, on the tourist: the influence of age, gender, project type and length of stay. Tour Manag Perspect 4:119–126

Bone J, Bone K (2018) Voluntourism as cartography of self: a Deleuzian analysis of a postgraduate visit to India. Tour Stud 18(2):177–193

Brightsmith DJ, Stronza A, Holle K (2008) Ecotourism, conservation biology, and volunteer tourism: a mutually beneficial triumvirate. Biol Conserv 141(11):2832–2842

Broad S, Jenkins J (2008) Gibbons in their midst? Conservation volunteers' motivations at the gibbon rehabilitation project, Phuket, Thailand. In: Lyons KD, Wearing S (eds) Journeys of discovery in volunteer tourism: international case study perspectives. CABI publishing, Wallingford, UK, pp 72–85. https://doi.org/10.1079/9781845933807.0072

Campbell R, Warner A (2016) Connecting the characteristics of international volunteer experiences with their impacts: a Canadian case study. Volunt Int J Volunt Nonprofit Org 27(2):549–573

Campera M, Budiadi B, Adinda A, Ahmad N, Balestri M, Hedger K, Imron MA, Manson S, Nijman V, Nekaris KAI (2021) Fostering a wildlife-friendly programme for sustainable coffee farming: the case of small-holder farmers in Indonesia. Land 10(2):121

Chapman CA, Corriveau A, Schoof VA, Twinomugisha D, Valenta K (2017) Long-term simian research sites: significance for theory and conservation. J Mammal 98(3):652–660

Chen LJ (2016) Intercultural interactions among different roles: A case study of an international volunteer tourism project in Shaanxi, China. Curr Issue Tour 19(5):458–476

Coren N, Gray T (2011) Commodification of volunteer tourism: a comparative study of volunteer tourists in Vietnam and in Thailand. Int J Tour Res 14(3):222–234

Cousins JA, Evans J, Sadler JP (2009) 'I've paid to observe lions, not map roads!'–an emotional journey with conservation volunteers in South Africa. Geoforum 40(6):1069–1080

Galley G, Clifton J (2004) The motivational and demographic characteristics of research ecotourists: operation Wallacea volunteers in Southeast Sulawesi, Indonesia. J Ecotour 3(1):69–82

Ghose T, Kassam M (2014) Motivations to volunteer among college students in India. Volunt Int J Volunt Nonprofit Org 25(1):28–45

Gray NJ, Campbell LM (2007) A decommodified experience? Exploring aesthetic, economic and ethical values for volunteer ecotourism in Costa Rica. J Sustain Tour 15(5):463–482

Liu TM, Leung KK (2019) Volunteer tourism, endangered species conservation, and aboriginal culture shock. Biodivers Conserv 28(1):115–129

Lorimer J (2009) International conservation volunteering from the UK: what does it contribute? Oryx 43(3):352–360

Nekaris KAI (2016) The little Fireface project: community conservation of Asia's slow lorises via ecology, education, and empowerment. In: Waller M (ed) *Ethnoprimatology. Developments in primatology: Progress and prospects.* Springer, Cham, pp 259–272. https://doi.org/10.1007/978-3-319-30469-4_14

Nugroho I, Pramukanto FH, Negara PD, Purnomowati W, Wulandari W (2016) Promoting the rural development through the ecotourism activities in Indonesia. Am J Tourism Manag 5(1):9–18

Oktavia AC, Mardiastuti A, Rahman DA (2020) Experience and the impact of voluntourism in Samboja Lestari Orangutan rehabilitation Center. IOP Confer Series Earth Environ Sci 528(1):012036

Otoo FE, Agyeiwaah E, Dayour F, Wireko-Gyebi S (2016) Volunteer tourists' length of stay in Ghana: influence of socio-demographic and trip attributes. Tour Plan Develop 13(4):409–426

Polus RC, Bidder C (2016) Volunteer tourists' motivation and satisfaction: a case of Batu Puteh village Kinabatangan Borneo. Procedia Soc Behav Sci 224:308–316

Russon AE, Susilo A (2014) Orangutan tourism and conservation: 35 years' experience. In: Russon AE, Wallis J (eds) Primate tourism: a tool for conservation. Cambridge University Press, Cambridge, UK, pp 76–97

Strzelecka M, Nisbett GS, Woosnam KM (2017) The hedonic nature of conservation volunteer travel. Tour Manag 63:417–425

Walpole MJ, Goodwin HJ (2000) Local economic impacts of dragon tourism in Indonesia. Ann Tour Res 27(3):559–576

Wearing S (2001) Volunteer tourism: experiences that make a difference. CABI Publishing, Wallingford, UK

Wieckardt CE, Koot S, Karimasari N (2020) Environmentality, green grabbing, and neoliberal conservation: the ambiguous role of ecotourism in the green life privatised nature reserve, Sumatra. Indonesia J Sustain Tourism. https://doi.org/10.1080/09669582.2020.1834564

Zahawi RA, Reid JL, Fagan ME (2020) Potential impacts of COVID-19 on tropical forest recovery. Biotropica 52(5):803

Chapter 3
Bukit Lawang and Beyond: Primates and Tourism from a Provider's Perspective

Andrea Molyneaux

Abstract The goal of this paper is to provide a unique perspective on tourism from the point of view of a business, ***Green Hill***, which has been providing conservation-focused tourism services in Bukit Lawang, North Sumatra, for 13 years. The business is owned by a local Sumatran and his UK-born wife who has a master's degree in primate conservation. Overall, the work and operation of Green Hill has had a positive impact on primate conservation and aligns with suggestions to improve the design of wild orangutan tourism and suggestions that it is preferable to market a more sustainable experience of searching for wildlife, focusing on the area and its flora and fauna, rather than guaranteeing a "photo op" of a primate at close range. Clear, practical guidance about implementation of principles of tourism along with long-term support must be given to providers. If primate tourism is to succeed and meet required goals, community-level local providers must be collaborated with on an equal footing, as neglecting to involve them is the most common cause of conflict and ultimately failure. It is essential to understand the area, the culture, the political history, the people, and the primates. The development process needs to be inclusive from the very beginning and community-level involvement is the key to success as they are the ones with the local knowledge. It is also crucial to understand the tourism market, the target audiences, and how to market services and manage expectations appropriately.

A. Molyneaux (✉)
Green Hill, Langkat, North Sumatra, Indonesia

Oxford Wildlife Trade Research Group, Oxford Brookes University,
Oxford, United Kingdom
e-mail: greenhillsumatra@googlemail.com

3.1   Introduction

Sumatra is the largest island in the Indonesian archipelago which consists of over 17,000 islands. Bukit Lawang is in the regent of Langkat in the province of North Sumatra where the population consists mainly of Acehnese, Batak Karonese, and Malay peoples. Historically, Sumatra was a land of forests split into many kingdoms and the economy was dominated by the lucrative spice trade which attracted colonizing European nations from the sixteenth century who monopolized the trade. Following a long colonial history, Indonesia gained independence in 1945. During the Dutch occupation in Sumatra, the Malay sultans and aristocracy were favored and given control of lands which, after independence in 1945 and a revolution against the sultans in 1946, were seized and over time have ended up becoming large estates, many of which are foreign owned (Islah 2011). While agriculture in North Sumatra is still a major contributor to the economy and focused on estate crops such as tobacco, rubber, and palm oil, since the 1960s destructive policies during the Suharto regime (1966–1998) and mismanagement of forests has led to a rapid growth of the timber industry and much deforestation (Barber 2002). While oil palm plantations surround Bukit Lawang, the local agricultural economy is based around crops such as rice, durian, betel nut, coconut, and banana. Commercial agriculture has shifted from cacao and rubber toward palm oil with many small farmers replacing rubber trees with palm oil as the price of the raw latex is not enough to buy 1 kg of rice. Tourism has a significant economic impact on the local area as indicated by permit numbers for Gunung Leuser National Park (2017: 11,067 foreign visitors, 5467 Indonesian visitors), and in the Langkat region, there are an increasing number of locations from which visitors can access nature-based tourism activities such as river swimming and jungle trekking: Simolap–Batu Katak–Bukit Lawang/Landak River–Tualang Gepang/Bukit Kencur–Simpang Dua–Batu Rongring–Tangkahan.

In terms of primates, Sumatra is best known for its two species of orangutans but Sumatra and its offshore islands are also home to four species of gibbon, four species of macaque, ten species of langur, two slow lorises, and a tarsier (Shepherd and Shepherd 2017). Despite this high level of primate diversity, wildlife tourism in North Sumatra is focused on orangutans at Bukit Lawang, elephants at Tangkahan (Gunung Leuser National Park), and wildlife in other popular destinations such as Berastagi and Lake Toba. In a wider nature context, Sumatra has much to offer including birding tours (Aceh), marine-based tourism (diving in Pulau Weh, surfing around the Mentawai Islands), and nature tourism in and around national parks (Kerinci Seblat, Bukit Barisan, Way Kambas). The small amount of published work on primate tourism in Sumatra has been done from the perspective of the primates (Russon and Susilo 2014; Dellatore et al. 2014; Ilham et al. 2017; Ilham et al. 2018; Barus et al. 2018) with other literature examining issues such as flagship species (Supriatna and Ario 2015), deforestation of primate habitats (Supriatna et al. 2017), and primate crop raiding (Marchal and Hill 2009). No literature exists from the perspectives of those working in the industry providing primate tourism services.

The goal of this paper is to provide a unique perspective on tourism from the point of view of a business, **Green Hill**, which has been providing conservation-focused tourism services in Bukit Lawang, North Sumatra, for 15 years. The business is owned by a local Sumatran and his UK-born wife who has a master's degree in primate conservation. First, I will give an overview of the history of Bukit Lawang and a summary of the primates of Bukit Lawang and how tourists may experience them. Then I will provide an overview of Green Hill, its practices and philosophy, and whether tourism in a remote area can be competitive. I will then use the data, information, and insights to reflect on whether tourism can have a positive impact, whether Green Hill has been successful, and where tourism providers might look for guidance. Finally, I will consider the lessons that can be learned and how the view from a provider in one local tourism hotspot has relevance for the wider debate on primate tourism in Indonesia and elsewhere.

A Note on Tourism Terminology

As a provider of tourism, one is faced with a myriad of different tourism terms which are often used interchangeably and without clear definitions. These include, but are not limited to: ecotourism, wildlife watching tourism, wildlife tourism, nature tourism, sustainable tourism, responsible tourism, and even a combination of terms (e.g., "responsustable tourism") (Mihalic 2016). This inconsistency has been shown to provide challenges for researchers evaluating and comparing primate tourism sites (Riley et al. 2015). Problems with the use and implementation of "ecotourism" have also been identified (Litchfield 2008; Russon and Wallis 2014; Macfie and Williamson 2010; Dekhili and Achabou 2015). From a provider's view point of view, it is challenging to know which terms apply and where one is to look for guidance on providing tourism services associated with primates, nature, and wildlife. Green Hill refers to the services they provide as "conservation focused tourism" and we urge researchers, conservationists, and all stakeholders to be consistent with any tourism terminology they use and to choose preferred terminology wisely.

3.1.1   The History of Tourism at Bukit Lawang

Bukit Lawang is a tourist-focused village in North Sumatra, Indonesia, situated along the banks of the Bohorok River on the border of Gunung Leuser National Park (GLNP). The village developed in response to the visitors to the Bukit Lawang (Bohorok) orangutan rehabilitation center. A brief history of the orangutan rehabilitation center and tourism in the area will give an important perspective on the current situation in Bukit Lawang (prior to the COVID-19 pandemic). The following summary was sourced from the 1997 book by Rijksen and Meijard, "Our Vanishing Relative," and supplemented with information from other sources (as indicated).

In 1971, Sumatra's first orangutan research and rehabilitation center was established in Ketambe, Aceh, by the Indonesian Government Nature Conservation Service (PPA) (now known as the Directorate General of Nature Resources and Ecosystem Conservation KSDAE) with financial support from WWF. In 1973, a second orangutan rehabilitation center was created with support from the Frankfurt Zoological Society (FZS) (and later WWF). This second center was established in a relatively inaccessible area of rainforest on the banks of the Bohorok river just within the border of the now Gunung Leuser National Park (GLNP). It was called the Bohorok orangutan rehabilitation center after the river and the small town of Bohorok 10 km away. The area is now known as the tourist village of Bukit Lawang.

For a number of years, both centers attracted visitors and the practice of regular close contact between staff and visitors with the rehabilitant orangutans raised the issue of anthropo-zoonotic disease transfer. In addition to affecting rehabilitant orangutans, the risk of disease transmission also threatened wild population of orangutans in the areas where rehabilitants were being released. Concerns were also raised that centers becoming tourist attractions might contradict the main objective of conservation. This concern led to the Ketambe center terminating rehabilitation in 1978 and shifting the focus solely to research. At this time, knowledge and research of orangutan biology was in its infancy and little was known about the most appropriate methods to rehabilitate orangutans without compromising wild populations. According to Rijksen and Meijaard, after these concerns were publicized in the late 1970s, support for orangutan conservation was withdrawn, and in 1980, WWF and FZS handed over administration of the center to the Indonesian authorities (Directorate General for Forest Protection and Nature Conservation—PHPA).

In the 1980s, rehabilitation and tourism continued at the BL/Bohorok site with visitors being allowed close interactions with rehabilitant orangutans. Guest houses began appearing that supported the burgeoning tourist activity of jungle trekking with some "private orangutan feedings" being offered. On April 23, 1991, the Director General of PHPA issued an instruction to close the rehabilitation centers at Bohorok (North Sumatra), Camp Leakey (Central Kalimantan), and Teluk Kaba (East Kalimantan), and in December 1991, at the Great Ape Conference in Jakarta, then President Suharto asked for international support to save orangutans. It is documented that WWF, who had previously been heavily involved with the rehabilitation centers, declined to offer any assistance stating that the concept of single species conservation was outdated, there was insufficient information about orangutan population status, and that "…sufficient resources were being spent in the conservation of the ecosystem of the ape" (Rijksen and Meijaard 1997, p. 141).

Reports of tourism at the BL/Bohorok center indicate that since the early 1990s the number of guest houses had increased greatly, visitors could buy a permit (for 4500 IDR) from the PHPA office to enter the national park and view the twice-daily orangutan feeding sessions, and the center was almost entirely supported by this tourism revenue (Eliot and Bickersteth 2000). After the rehabilitation activities were officially terminated in 1994, the viewing platform and twice-daily feedings (still open to the public) were continued by PHPA to support rehabilitant orangutans that still visited the feeding platform and to try and limit the "private feedings" during jungle treks. Limited facilities were maintained to deal with some rehabilitant orangutans that were problematic and some that crossed the river to the guest houses.

Although the orangutan feeding platform and viewing sessions were discontinued around 2015 and the price of national park permits was increased from 20,000 IDR to 150,000 IDR, there continues to be a thriving tourism industry focused on jungle trekking activities in the national park. While the problems of close contact and tourism were identified in the early 1970s, the problem persists in 2020 despite Indonesian law forbidding harm to orangutans (UU No5 1990) and rules and guidelines from the national park authorities regarding wildlife viewing and safety information (Green Hill 2020). Research using data collected from Instagram has revealed non-compliance by visitors with the rule to keep a minimum distance of 10 meters from orangutans. Major concerns have been expressed that without widespread awareness of this issue and appropriate actions by all stakeholders, there is a very real risk of anthropo-zoonotic disease transfer, which includes COVID-19, from human visitors to the orangutan population in Bukit Lawang (Molyneaux et al. 2021).

3.1.2 Brief Overview of the Current Tourism Market

An analysis of national park permit numbers from August 2019 to February 2020 (Fig. 3.1) revealed a total of 6819 foreign permits and 5453 domestic permits. The monthly variance shows foreign visitors peak in the usual high season of July–September and domestic (Indonesian) visitors peak in December and January.

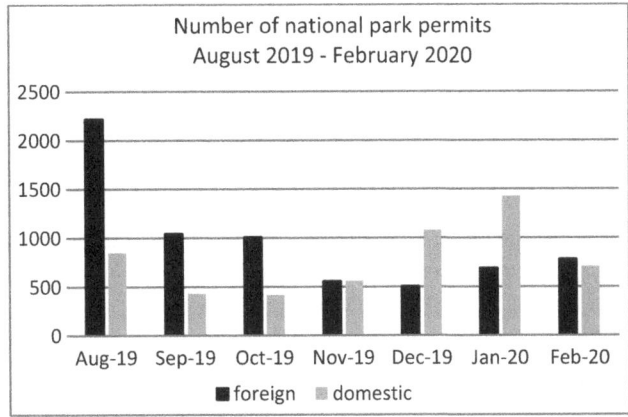

Fig. 3.1 Number of National Park Permits from August 2019 through February 2020 broken down according to domestic vs. international tourists

Analysis of visitor nationality from Green Hill trek data for April 2019 to March 2020 revealed the most common nationalities were UK (18%) and Holland (17%). Australia and Germany represented 11% each, USA 6%, Sweden, Switzerland, Belgium, and Canada 4% each, Denmark and Poland 3% each, New Zealand, Malaysia, and Japan 1% each.

3.1.3   The Primates of Bukit Lawang and beyond

With the history of the Bohorok rehabilitation center and feeding platform, tourism in the area is focused on jungle trekking into GLNP to search for some of the completely free ranging rehabilitant orangutans which are entirely habituated and very used to human presence. By focusing on one species, it can be thought of as a monoculture of tourism. Despite the fact that Sumatra is a biodiversity hotspot and little attention is given to promoting awareness of other species that can be seen in and around Bukit Lawang. A citizen science project registered as part of the United Nations Decade of Biodiversity analyzed wildlife photographs and revealed 314 different species. Guests in and around the national park report seeing noteworthy species such as pangolin, colugo, and binturong, and a very recent camera trap on private land near the national park border revealed a Sumatran serow! In terms of primates, there are eight species in the immediate area and a ninth species has been observed 2 h away near the Simolap hotsprings (Shepherd and Shepherd 2017). The diversity of primates to be viewed in this area is illustrated in Table 3.1.

3.2   Overview of Green Hill Practices and Philosophy

Green Hill is a conservation-focused tourism provider based in Bukit Lawang that also operates a long running environmental education and community conservation program. The owners are a mixed nationality couple who bring unique experiences and skills to the provision of a tourism service. Mbra was born in Bukit Kencur just 30 min from Bukit Lawang and thus is part of and has intimate knowledge of the local community. He has been involved in tourism as a forest guide for over 20 years, many of which were with a scientific research project and is now focusing on the management of Kuta Langis Ecolodge and farm. Andrea was born in the UK where for almost 20 years she worked for as a biomedical scientist and a senior scientist for the UK government. Following long-term volunteering on conservation projects in both Kalimantan and Sumatra, she has now lived in Sumatra for over 12 years, has spent a lot of time trekking in remote areas of rainforest, has a master's degree in Primate Conservation, and is working toward a PhD in conservation, tourism, and education. The cornerstone of their conservation work is a children's nature club and community library in the remote village of Tualang Gepang. Through these efforts, they provide enhanced creative education anchored to the UK and Indonesian

Table 3.1 Primates observed in Bukit Lawang and their conservation status

Type of organism	IUCN red list status	Where they are seen [in the BL area]
Sumatran orangutan *Pongo abelli*	Critically Endangered	Standard BL trails: Occasionally wild orangutans are seen but commonly encountered are rehabilitants, especially Jecky and Mina, who are extremely habituated and mostly on the ground. Sometimes seen at the trekking campsites upriver in BL
Siamang *Symphalangus syndactylus*	Endangered	Infrequently seen on longer trails in BL area. Some groups habituated and have been known to raid picnic lunches
White handed gibbon *Hylobates lar*	Endangered	Occasionally seen high in the canopy on the standard BL trails but no known instances getting close or of becoming habituated
Thomas langur *Presbytis thomasi*	Vulnerable	Often seen in the rubber trees on border of entrance to GLNP and in the earlier parts of the standard BL trails. Very habituated and will take fruit from visitors. Also often seen in the upriver area in forest behind guest houses but they are not habituated
Silver langur *Trachypithecus cristatus*	Vulnerable	Never seen on standard BL trails. Most often seen in the secondary border habitat before the entrance to GLNP. Not habituated at all
Long tailed macaque *Macaca fascicularis*	Endangered	Frequently seen in the rubber trees on border of entrance to GLNP, in the earlier parts of the standard BL trails, along the riverside, in the forest close to guest houses, camping ground. They are habituated and often raid shops and restaurants if they get the chance. Local tourists have been known to feed them by camping ground
Southern pig-tailed macaque *Macaca nemestrina*	Endangered	Rarely seen on the standard BL trails and always lone males. Infrequently a group has been seen in the upriver area in forest behind guest houses
Sunda slow Loris *Nycticebus coucang*	Endangered	Never seen by guests due to its nocturnal nature. Reports of sightings by local residents on electricity cables at camping ground BL and at Simolap HotSprings where sightings have been reported more often and possible hunting
Black Sumatran langur *Presbytis sumatrana*	Endangered	Small group spotted once only in a rural farmland/ secondary forest area near the Saringgana waterfall. Thought to be at the northernmost edge of its range

schools' national curriculum for science. They aim to inspire and empower the children to learn about and protect their own environment and the associated wildlife. The conservation work is self-funded by their ethical tourism program which specializes in jungle treks in areas off the beaten track and well away from the busy tourist trails.

Green Hill encompasses two properties: Green Hill Guest House and the more recently established Kuta Langis Ecolodge. Green Hill Guest House was created in

2007 in the upriver area of Bukit Lawang set back from the Bohorok River and on the site of a derelict guest house which had survived the 2003 flood. The Green Hill Guest House was established with the purpose of operating a conservation-focused and environmentally sympathetic business on a small scale (five rooms). In 2015, Green Hill purchased four hectares of palm oil plantations, 20 min from Bukit Lawang, between the villages of Tualang Gepang and Bukit Kencur. The land extends from the village down a hillside and connects directly to the border of GLNP by the Kerapoh River. The majority of the palm oil was felled and left to rot into the ground. The land has since been successfully replanted and regenerated with a mixed crop of durian and other fruit trees in some areas, while other areas have been left to "rewild." Kuta Langis Ecolodge was created in 2018 using sustainably sourced materials to provide the opportunity for guests to stay in this seldom visited area and to experience community-based tourism. The ethos here is to go off grid, because while there is electricity at the site, the water source is a natural spring, and there is no Wi-Fi. Unlike Bukit Lawang, there is no tourist infrastructure such as souvenir stalls, restaurants, and other guest houses. When required, the Ecolodge is staffed by Green Hills trekking teams and their family members from the local village.

3.2.1   Green Hill Jungle Trekking Philosophy and Rule Adherence

Green Hill specializes in offering trekking in remote areas of the national park away from the standard Bukit Lawang trails. The aim here is to reduce pressure on the standard trails and campsites in Bukit Lawang (which get very busy in high season and have negative impacts) and to promote trekking that does not focus on seeing orangutans. A popular option is the two-day Discovery Trek which combines trekking in two different areas and camping at a pristine location within the remote area. For over ten years, Green Hill has been the only operator routinely offering treks in the geographically close but difficult to access area of jungle near the villages of Tualang Gepang/Bukit Kencur which is virtually untouched national park rainforest. Green Hill employs a growing team of people from the local villages who also follow the rules and regulations and have become ambassadors for sustainable trekking behaviors. On longer treks, they have unique routes which take guests from the Bukit Lawang trails through to Tualang Gepang/Bukit Kencur and on to areas of remote jungle near the villages of Simpang Dua and Batu Rongring. In 2019, Green Hill began working with a local community conservation group and a local fledgling tour company just 2 h from Bukit Lawang to take guests trekking in the national park near their villages of Kinankong and Simolap. Information from the Green Hill website ((A) Appendix 1) illustrates how these alternative trekking areas are promoted and how guests' expectations managed. The website and social media channels are extremely careful not to guarantee any specific wildlife sightings, and guides and staff are advised how to interact with guests. For instance, instead of just talking about orangutans or asking, "How many orangutans did you see?" it can be beneficial to ask, "Did you see a lot of wildlife?" or "Did you enjoy being in the rainforest?"

There are negative associations with Bukit Lawang related to the uncontrolled manner of tourism, its negative impact on the health of the orangutans, and poor management (Kuze et al. 2011; Dellatore et al. 2009; Susilawati et al. 2020). Aware of these negative aspects and surprised by the lack guidance for guides, visitors, and tourism providers, Green Hill decided to take action. In 2007–2008, they produced a set of detailed rules for trekking based on published literature and accounts from primate tourism sites. These rules were distributed widely and shared with the local guide association (HPI) and non-governmental organizations that were active in the area at that time. Following the death of two infant orangutans in 2018, Green Hill began intense liaison with national park authorities and together have developed and implemented a detailed education and communication campaign to promote awareness of safe trekking practices (Green Hill 2020). The campaign resulted in the following:

- A metal signboard with the rules has been permanently placed at the entrance to the national park.
- A detailed trifold brochure outlining the rules and their background.
- A poster/flyers and selfie booth explaining why close selfies with animals are bad (in conjunction with AWCP and IUCN primate human contact group).
- Updated trekking rules (based on IUCN great ape health monitoring guidelines) and provided advice to stakeholders regarding the impact of COVID-19.
- COVID-19-specific posters/flyers and stickers.
- "Mobile Education Conservation Units": 40+ plastic rain canopies for becaks with COVID-19-specific information.

The rules underpin the trekking operations at Green Hill and all guides, assistants, and staff must follow them strictly. In order to engage with local forest guides, to promote awareness of the trekking rules, and to provide a foundation in basic science, biology, and other relevant topics, Green Hill has recently launched an open access social learning page on Facebook. To frame their work on the issue, Green Hill created a long running campaign entitled **KEEP WILDLIFE WILD** and has included materials on their website, on their social media channels, and on display in the guest house. Green Hill has been extremely active in promoting awareness of these issues, and in creating and maintaining a community science project to raise awareness about inappropriate selfies on Instagram (Molyneaux et al. 2021). Another key component of the Green Hill philosophy is to promote the awareness of biodiversity in the area using the hashtag **#morethanorangutan,** since other species are often overlooked.

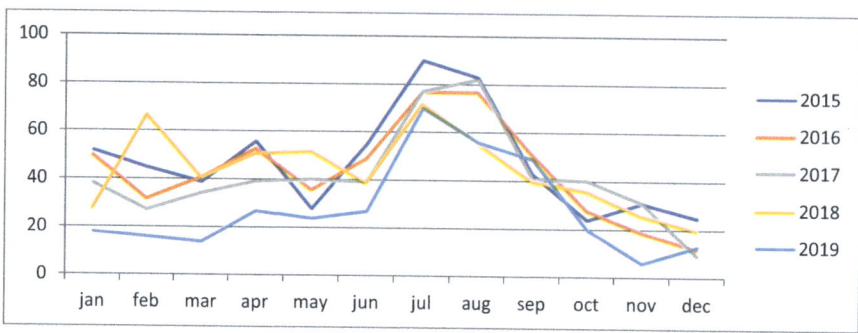

Fig. 3.2 % Occupancy rates at Green Hill per month 2015–2019

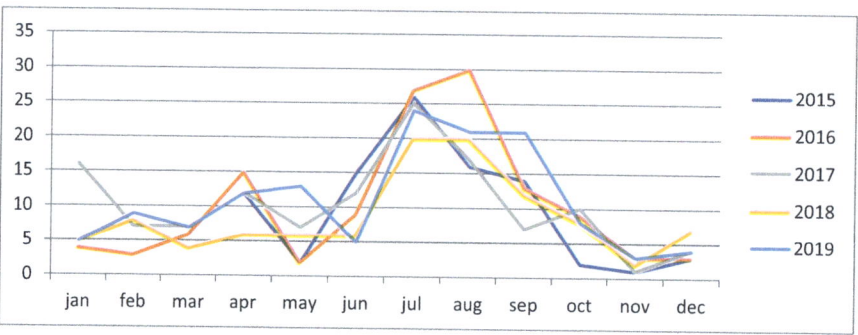

Fig. 3.3 Number of Treks booked per month 2015–2019

3.2.1.1 Jungle Trekking in Remote Areas

Data generated by calculating monthly occupancy rates (Fig. 3.2) and number of treks (booked via email: Fig. 3.3) for 2015–2019 illustrate that occupancy rates and trek numbers at Green Hill follow a consistent seasonal pattern. While occupancy rates have decreased, the number of treks taken has increased. This indicates that Green Hill remains competitive, that discerning visitors are interested in such tourism services and thus the conservation-focused approach is successful and is competitive in the current busy market. It is extremely important to remain competitive in order to continue generating income for conservation work and that promotion and appropriate marketing effort toward different types of consumers need to be maintained (Buckley and Mossaz 2018).

To investigate whether visitors to the Bukit Lawang would be interested in a more naturalistic and sustainable approach to trekking focused on the community and environment in an area with no rehabilitated orangutans and thus no guaranteed sightings, the jungle trekking data were analyzed for the number, type, location, and duration of treks. In terms of location when comparing total numbers of treks purely in Bukit Lawang with all other treks (combination of or different locations) over 5 years only 36% of treks were in Bukit Lawang and 64% were a combination or

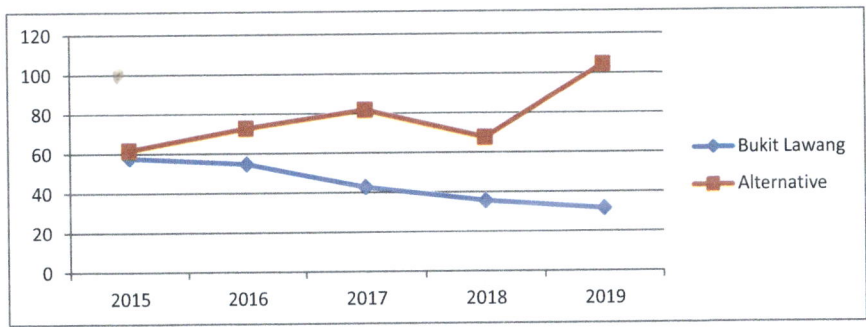

Fig. 3.4 A comparison of the number of treks in Bukit Lawang with alternative locations 2015–2019

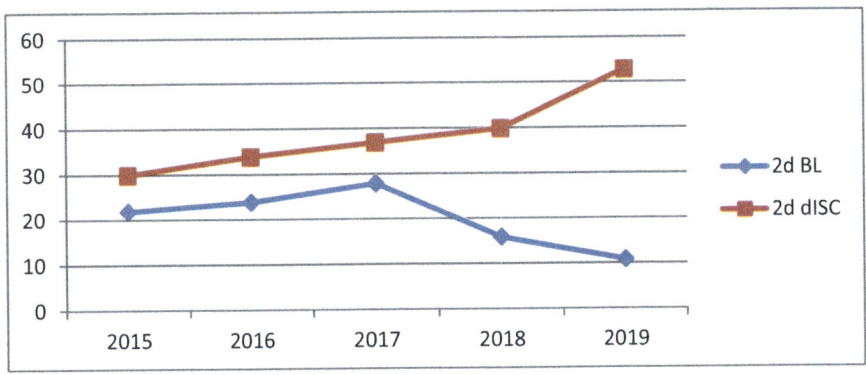

Fig. 3.5 A comparison of 2-day treks in Bukit Lawang with the 2-day Discovery Trek 2015–2019

completely different location which indicates successful marketing of the remote and new area for trekking (Fig. 3.4).

Two-day treks remain the most popular option with numbers remaining stable and representing 50% of all treks over the 5-year period. Directly comparing two-day Bukit Lawang and Discovery Treks (Fig. 3.5) reveals that Discovery Treks have always been more popular and the difference has become increasingly polarized with Discovery Treks far outweighing ones in purely Bukit Lawang.

For three-day or longer treks, Discovery Treks outstrip Bukit Lawang treks representing 97% of treks. Guests are increasingly interested in staying at the Kuta Langis Ecolodge (even without specific marketing yet) which indicates that visitors are definitely interested in getting away from busy trails and having a more naturalistic experience. In 2019, Green Hill began a trial of offering day trips and longer experiences in an even more remote and inaccessible area of GLNP, Simolap, and initial findings are extremely positive.

Can Tourism Have Positive Impact on Primate Conservation? The commonly held assumption that ecotourism benefits conservation has been called into question with suggestions that it is less sustainable in Asia (Krüger 2005), that it may not achieve conservation goals (Higham 2007), and that threatened species may suffer negative impacts from ecotourism (Buckley et al. 2016). Potential disease transmission from tourists to primates has been identified as a major issue of concern (Köndgen et al. 2008; Litchfield 2008; Muehlenbein and Wallis 2014; Dunay et al. 2018; Hanes et al. 2018; Weber et al. 2020) and especially so in recent times due to the COVID-19 pandemic (Gillespie and Leendertz 2020, Lappan et al. 2020, Santos et al. 2020, Glasser et al. 2021, Molyneaux et al. 2021). However, it is believed that tourism can have a positive impact on primates if managed correctly (Desmond and Desmond 2014; Kurita 2014; Muehlenbein and Wallis 2014), especially if developed in conjunction with researchers and conservationists (Russon and Wallis 2014). Using population viability analysis to calculate the net effects of ecotourism revealed that for orangutans, zero or low levels of tourism could lead to extinction whereas medium to high levels of ecotourism had an overall positive effect that offset the impacts of logging. (Buckley et al. 2016). Bukit Lawang and its neighbor, Tangkahan, in which elephant-focused tourism operates, are in the Langkat region of North Sumatra, 80 km from the area of Besitang which has a high level of forest encroachment and unresolved conflict (Purwanto 2016). The combined presence of long-term local providers of tourism activities and infrastructure at both locations acts as a deterrent to encroachment and can be viewed as having a positive impact on primate conservation in that it protects the forest habitat. (Russon and Russell 2005; Dellatore et al. 2014).

3.2.1.2 Benefits of Promoting New Alternative Tourism Areas

The success of Green Hill in promoting a new area for trekking and tourism bodes well for the future of Bukit Lawang and tourism in the Langkat area of North Sumatra. It demonstrates that other areas can be promoted with the right marketing and knowledge, thereby taking some of pressure off the busy Bukit Lawang trails and wildlife. This will also help spread some income to the remoter areas and potentially help protect their forest and wildlife. Another benefit of including these areas would be to increase awareness of biodiversity and to promote the area as a nature experience destination. As Green Hill is part of the local community, they have been able to contribute (e.g., income, education, and outreach) in ways not possible with the conventional model of conservation projects. Following informal discussions with village trekking guides and other community members, Green Hill donated a plot of land to be used as a graveyard. Working in conjunction with local government, another plot of land was donated on which a village hall and meeting room was built.

As remote areas are inherently less visited and are thus more susceptible to influences such as illegal hunting and logging, trekking in the remote areas can act as a deterrent, as does having a permanent physical presence in the form of the Kuta Langis Ecolodge on the border of the national park. When Green Hill initially started trekking in this area ten years ago, it was quite common to find snare traps but now they are seldom found. Green Hill's local guides believe it is because they take people trekking there. Additionally, the long-term community conservation and outreach work of Green Hill is believed to have a positive impact as evidenced by the fact that there are no longer pet macaques in the Nature Club village and that community members now self-manage or report incidences of potential conflict with orangutans to Green Hill, who facilitates liaison with local national park authorities.

Overall, the work and operation of Green Hill has had a positive impact on primate conservation and aligns with suggestions to improve the design of wild orangutan tourism (Russon and Susilo 2014) and suggestions that it is preferable to market a more sustainable experience of searching for wildlife, focusing on the area and its flora and fauna, rather than guaranteeing a "photo op" at close range (Macfie and Williamson 2010; Weber et al. 2020).

Where Can Service Providers Look for Guidance on Primate Tourism? Primate tourism developed in conjunction with researchers and conservationists who have long-term commitments to the area and remain involved is more likely to have a positive impact (Russon and Wallis 2014). As a tourism provider in a primate habitat country looking for advice and guidance to operate in a conservation-focused manner, one could logically look toward those concerned with primate conservation and publishing literature on the subject as the ones to inform such initiatives as they often advise on actions to be taken by providers.

Orangutans

Well-designed education programs are seen as a way to mitigate the threat of species extinction and to promote positive change (Brown et al. 2019; Weber et al. 2020). It has been advised that educational programs are needed for guides, visitors, and local residents in Bukit Lawang and that national park authorities should be active in enforcing trekking guidelines (Dellatore et al. 2014). Perhaps idealistically, Russon and Susilo (2014) advised that tourism with rehabilitated orangutans should be stopped and the management of orangutan tourism should be designed to prevent problems and limit damage (Russon and Susilo 2014). Best practice tourism guidelines from the IUCN (Macfie and Williamson 2010), specifically in relation to orangutans, state: "Tourism managers should impose rules to stop the feeding of free ranging orangutans by both tourists and guides, and indeed prohibit the carrying of any food into the forest." Additionally, in an appendix, they have a set of guidelines for visitors to Bukit Lawang which mirrors those written by Green Hill. In Bukit Lawang, other than the long-term work of Green Hill and a well-designed but unsustained education program for guides by a group of NGOs in 2008–2009, little sustained action has been implemented to mitigate negative impacts of tourism. Lack of guidance from local government to the community is cited as one of the main problems in developing ecotourism in Tanjung Putting National Park, which is home to orangutan viewing tourism at a historic research site, Camp Leakey (Meilida and Tuah 2020). This is likely to be a contributory factor in Bukit Lawang.

Other Primate Tourism Sites in Indonesia While there are ~39 species of primates in Indonesia, there is insufficient literature on primate-focused tourism and, other than orangutans (in Sumatra and Kalimantan) and macaques in Bali, it is perhaps an undeveloped area of research focus. In Sumatra and its offshore islands with ~19 species of primates, 9 of which can be seen in and around Bukit Lawang area, little attention is given to anything other than orangutans. The difficult to reach Mentawai Islands off the west coast of Sumatra, best known for surfing and tribe tourism, have four endemic primate species. A conservation masterplan in 1980 recommended development of ecotourism in Siberut National Park (WWF 1980), and a revised conservation action plan in 2006 suggested that formally protecting the Peleonan forest could allow opportunities for ecotourism development (Whittaker 2006). The Javan Silvery Gibbon (*Hylobates moloch*) is one of five primates that remain on the island of Java (Supriatna et al. 2010) and it has been suggested that the development of ecotourism utilizing its "charm" could generate income for local stakeholders (Supriatna 2006). It would appear that these areas of potential have remained undeveloped as other than a fledgling tourism project from an Indonesian NGO (SwaraOwa 2021) there is little information available on primate watching tourism in the Mentawai Islands. Another Indonesian-led research project, "Owa Halimun," in west Java is in the early stages of developing gibbon tourism with local stakeholders. However, this has been put on hold due to the COVID-19 pandemic. (Roktavini A pers. comm. 2021) On the island of Sulawesi, all 17 species of primates are endemic and threatened by loss of forest habitat (Supriatna et al. 2020). Studies have identified negative impacts of tourism on the black crested macaque in Tangkoko Nature Reserve (Paulsen 2009; Kinnaird and O'Brien 1996) and a research/conservation project established in 2007 is developing an ecotourism program integrated with local communities, guides, and government agency (Selamatkan Yaki 2020).

Primate Tourism in a Wider Context In relation to gorilla tourism, it has been advised that studies to monitor and advise ecotourism management strategies should be carried out (Shutt et al. 2014), that guidelines should be provided to tourists before arrival, and that there should be widespread awareness of the issues and enhanced educational interpretation (Litchfield 2008). Research at a Barbary macaque site in Morocco advises that general disease prevention strategies aimed at reducing opportunities for contact between tourists and macaques should be adopted and that the IUCN guidelines on great ape tourism be extended and enforced in relation to other primates (Carne et al. 2017). Another macaque study in Japan advised that the management of free ranging monkey parks take a leading role in education and enforcement of appropriate behaviors regarding safe distances (Kurita 2014).

This brief review illustrates that tourism is often mentioned in published literature and actions for providers are given as potential solutions, but rarely is any practical advice given to providers of tourism services, and primate tourism in Indonesia is under developed. Implementation of ecotourism is complex, as is compliance with guidelines, which receives little attention in conservation literature despite it underpinning much of nature conservation (Keane et al. 2008; Arias 2015; Solomon et al. 2015; Fairbrass et al. 2016). There does seem to be a missing link joining the work and advice of researchers with the providers of the tourism services.

> **Key Points that Have Enabled Green Hill to Have Positive Impact on Primates**
>
> - Possessing in-depth knowledge and experience of primate biology and conservation.
> - Promoting knowledge and awareness of the high levels of biodiversity that the area has to offer, rather than focusing on a single species form of tourism.
> - Using a long-term, consistent, and sustainable approach which is collaborative.
> - Demonstrating genuine respect for and collaborate with the local community in more than just a cursory manner.
> - Valuing knowledge, understanding, and appreciation of the area, its history, and the community.
> - Managing guests' expectations by giving clear and consistent information at every step of the customer journey.
> - Having clear rules and expectations and facilitating understanding rather than blindly expecting compliance.
> - Providing education and sharing knowledge in a variety of formats for different audiences (guests, guides, community, etc.) that reflects different learning styles.

3.3   Conclusions

Green Hill is a successful provider of self-funded and self-sustaining conservation-focused tourism services in one local tourism hotspot. They have successfully extended that service to a remote area using a more sustainable approach that focuses on the nature experience rather than guaranteed sightings of a single primate species. The owners of Green Hill have unique backgrounds and in-depth knowledge and experience of the area and its wildlife. Additionally, they are a respected part of the community with whom they work with and have demonstrated long-term commitment. This collaborative approach has resulted in a sustained positive impact that benefits the local human and primate populations. Lastly, they strictly manage visitor expectations at all stages of the journey by using clear terminology and descriptions of their services.

We suggest that for primate tourism to be effective in other areas in Indonesia and beyond, primates should not be the sole focus of a tourism program, but part of a whole experience which also includes the wider biodiversity and ecology of the area. Clear, practical guidance about implementation of principles of tourism along with long-term support must be given to providers. Tourism, be it "eco" or otherwise, should not be seen as a panacea and there is not a "one size fits all" model or set of guidelines that can be applied equally worldwide. If primate tourism is to succeed and meet required goals, community-level local providers must be collaborated with on an equal footing, as neglecting to involve them is the most common cause of conflict and ultimately failure (Hawkins 2004). It is essential to understand the area, the culture, the political history, the people, and the primates, and the development process needs to be inclusive from the very beginning and community-level involvement is the key to success as they are the ones with the local knowledge. It is also crucial to understand the tourism market, the target audiences, and how to market services and manage expectations appropriately.

Researchers have identified the value of appropriately implemented and managed tourism, but need to learn from successful providers and projects, and need to use clear and appropriate terminology. Such stakeholders should work together to come up with practical solutions to identified problems that are reasonably priced and or do not require long-term funding. It is possible that conservation-focused, small-scale, community-level tourism could become self-sufficient, thus perpetuating the positive effects and reducing the reliance of conservation on external funding. Small-scale local providers, working with appropriate support and guidance from researchers and conservationists, and in coordination with local government, could be a key factor in the future of primate conservation.

Appendix 1

From Website www.greenhillbukitlawang.com

Bukit Lawang trails represent only a very small part of the Gunung Leuser National Park and this area is reasonably well protected due to the presence of the tourism village and its frequent visitors. There are many other stunning areas in GLNP and the Leuser Ecosystem which are virtually untouched and home to incredible wildlife but are less protected and thus more vulnerable to human impact (hunting, poaching, extraction of natural resources...). Why not get the best of both worlds by joining us for a trek that combines trekking in two or more different areas of rainforest....see our DISCOVERY TREK options. When you join us for a trek in the less visited areas you will be helping us to continue our conservation work in the jungle and the surrounding community's. Our presence in the remote village and taking guests trekking in surrounding jungle has had an impact as our tracking guides/community rangers have reported a great decrease in the numbers of snare traps and bird hunting areas.

Trekking in Bukit Lawang:

The scenery here is stunning, trails are well established and the wildlife is quite habituated so this means that you have quite high chances of spotting wildlife which includes orangutans, thomas leaf monkeys and great argus pheasants etc. The trails are well trekked and in high season can be very busy; as a result the wildlife does not exhibit normal behaviours (e.g. often coming down to the ground etc). A big draw about trekking here is that you have the option of returning after the trek by river rafting.

    TREKKING IN BUKIT KENCUR - TUALANG GEPANG....AND BEYOND. ...is 30 min away by motorbike and really is a virtually untouched jungle. The environment is truly amazing and breathtaking. The wildlife here shows 100% natural behaviour and is more difficult to spot...but it is definitely there (see our facebook page) so the focus here is on experiencing a pristine rainforest environment away from the tourist trails. This is a real ECOTREK and conservation option as we have a team of tracking guides from the local villages, some who used to be hunters and by trekking with us they have an alternative income. They also act as community rangers and inform national park authorities of any developments such as fallen trees blocking river flow etc.

Appendix 2

Table 3.A.1 Occupancy rates at green hill and number of treks per month 2015–2019

	2015		2016		2017		2018		2019	
	OR %	treks	OR %	treks	OR %	treks	OR %	treks	OR %	Treks
Jan	52	5	50	4	38	16	28	5	18	5
Feb	45	9	32	3	27	7	67	8	16	9
Mar	39	7	41	6	34	7	41	4	14	7
Apr	56	12	53	15	39	12	51	6	27	12
May	28	2	36	2	40	7	52	6	24	13
Jun	55	15	49	9	39	12	39	6	27	5
Jul	90	26	77	27	77	25	72	20	70	24
Aug	83	16	77	30	82	17	56	20	56	21
Sep	43	14	51	13	41	7	40	12	49	21
Oct	24	2	28	9	40	10	36	8	20	8
Nov	31	1	19	3	31	1	26	2	6	3
Dec	25	3	12	3	9	4	20	7	13	4
Ecolodge	2		2		10		14		21	
Av	48	10	26	10	42	10	44	9	24	11
Total		117		124		125		104		132

OR % occupancy rate at Green Hill guest house in Bukit Lawang; *KLEL* no of times guests stayed at Kuta Langis Ecolodge (as part of the Discovery Trek)

Appendix 3

Table 3.A.2 Number, type, location, and duration(days) of treks 2015–2019

	Bukit lawang rafting					2/2	Discovery trek						klel	TG/BK		SIM
	½	1	2	3	4		2	3	4	5	6	7	1	2		
2015	7	28	22		1	10	30	12	4		2		2	3	1	
2016	4	22	24	1		7	34	22	3				2	3	4	
2017		15	28			9	37	23	13				10			
2018	4	15	16	1		8	40	14	4	2			14			
2019	6	14	11	1		3	53	24	9	6		1	21	2		7

The following definitions were used:

Bukit Lawang: treks along the standard trails in GLNP close to the tourist village of Bukit Lawang and guests return to the village by rafting along the Bohorok river on rafts made from inner tubes. Campsites for overnight treks are mainly up river from the village, can be reached in an hours walk upriver or by walking through the jungle for approximately 5–6 h.

TG/BK: treks in GLNP close to the villages of Tualang Gepang and Bukit Kencur.(TG/BK).

2/2 trek: two different one day treks in the two different locations with the guests sleeping in Green Hill.

The **Discovery Trek:** is a combination of the two locations, for 2 days or longer with overnight camping in the remote jungle area location at a select few pristine locations OR staying at the remote Kuta Langis Ecolodge. (KLEL).

Simolap (sim): day trips or longer to the nat park 2 h away near village of Simolap.

References

Arias A (2015) Understanding and managing compliance in the nature conservation context. J Environ Manag 153:134–143

Barber C (2002) Conserving the peace: resources, livelihoods and security. In: Matthew R, Halle M, Switzer J (eds) Canada, international institute for sustainable development and IUCN

Barus JA, Hidayat JW, Maryono M (2018) Primates (Symphalangus syndactylus syndactylus, Macaca nemestrina, Macaca fascicularis) population in the Ape Park tourist area forest for special purpose of Aek Nauli. E3S Web Conf. 73(2018):04018

Brown ER et al (2019) Testing efficacy of a multi-site environmental education programme in a demographically and biologically diverse setting. Environmental Conservation:1–7. https://doi.org/10.1017/S0376892919000389

Buckley RC, Morrison C, Castley JG (2016) Net effects of ecotourism on threatened SpeciesSurvival. PLoS One 11(2):e0147988

Buckley R, Mossaz A (2018) Private conservation funding from wildlife tourism enterprises in sub-Saharan Africa: conservation marketing beliefs and practices. Biol Conserv 218(2018):57–63

Carne C, Semple S, MacLarnon A et al (2017) Implications of tourist–macaque interactions for disease transmission. EcoHealth 14:704–717. https://doi.org/10.1007/s10393-017-1284-3

Dekhili S, Achabou M (2015) The perception of ecotourism. Semantic profusion and tourists' expectations [*]. RIMHE: Revue Interdisciplinaire Manage Homme Entre 5(5):3–20. https://doi.org/10.3917/rimhe.019.0003

Dellatore DF, Waitt CD, Foitova I (2009) Two cases of mother–infant cannibalism in orangutans. Primates 50:277–281

Dellatore DF, Waitt CD, Foitova I (2014) In: Russon AE, Wallis J (eds) Primate tourism: a tool for conservation? Cambridge University Press

Desmond J, Desmond J (2014) Evaluating the effectiveness of chimpanzee tourism. In: Russon A, Wallis J (eds) Primate tourism: a tool for conservation? Cambridge University Press, Cambridge, pp 199–212. https://doi.org/10.1017/CBO9781139087407.014

Dunay E, Apakupakul K, Leard S, Palmer JL, Deem SL (2018) Pathogen transmission from humans to great apes is a growing threat to Primate Conservation. EcoHealth 15(1):148–162. https://doi.org/10.1007/s10393-017-1306-1. Epub 2018 Jan 23

Eliot J, Bickersteth J (2000) Footprint Sumatra handbook: the travel guide. Passport Books, UK

Fairbrass A, Nuno A, Bunnefeld N et al (2016) Investigating determinants of compliance with wildlife protection laws: bird persecution in Portugal. Eur J Wildl Res 62:93–101

Gillespie TR, Leendertz FH (2020) COVID-19: protect great apes during human pandemics. Nature 579:487

Glasser DB, Goldberg TL, Guma N et al (2021) Opportunities for respiratory disease transmission from people to chimpanzees at an east African tourism site. Am J Primatol 83:e23228. https://doi.org/10.1002/ajp.23228

Green Hill (2020) Wildlife watching leaflet and safe selfie poster. Green Hill website. https://www.greenhillbukitlawang.com/green-hill-and-conservation

Hawkins DE * (2004) Sustainable tourism competitiveness clusters: application to World Heritage sites network development in Indonesia. Asia Pac J Tour Res 9(3):293–307, https://doi.org/10.1080/1094166042000290682

Hanes AC, Kalema-zikusoka G, Svensson MS, Hill CM (2018) Assessment of health risks posed by tourists visiting mountain gorillas in Bwindi impenetrable National Park, Uganda. Primate Conservation 32:123–132

Higham J (2007) Critical issues in ecotourism: understanding a complex tourism phenomenon. (1st ed.). Routledge. p1–19.

Ilham K, Rizaldi N, J. & Tsuji, Y. (2017) Status of urban populations of the long-tailed macaque (Macaca fascicularis) in West Sumatra, Indonesia. Primates 58:295–305

Ilham KR, Nurdin J, Tsuj Y (2018) Effect of provisioning on the temporal variation in the activity budget of urban long-tailed macaques (Macaca fascicularis) in West Sumatra, Indonesia. Folia Primatol 89(5):347–356

Islah M (2011) Half of Indonesia's oil palm plantations foreign-owned. Eco-business. https://www.eco-business.com/news/half-of-indonesias-oil-palm-plantations-foreign-owned/. Accessed 25 Mar 2021.

Keane A, Jones JPG, Edwards-Jones G, Milner-Gulland EJ (2008) The sleeping policeman: understanding issues of enforcement and compliance in conservation. Anim Conserv 11:75–82

Kinnaird M, O'Brien T (1996) Ecotourism in the Tangkoko DuaSudara nature reserve: opening Pandora' box? Oryx 30(1):65–73. https://doi.org/10.1017/S0030605300021402

Köndgen S, Kühl H, N'Goran PK, Walsh PD, Schenk S, Ernst N et al (2008) Pandemic human viruses cause decline of endangered great apes. Curr Biol 18(4):260–264

Krüger O (2005) The role of ecotourism in conservation: panacea or Pandora's box? Biodivers Conserv 14:579–600

Kurita H (2014) Provisioning and tourism in free-ranging Japanese macaques. In: Russon A, Wallis J (eds) Primate tourism: a tool for conservation? Cambridge University Press, Cambridge, pp 44–55. https://doi.org/10.1017/CBO9781139087407.005

Kuze N, Dellatore D, Banes G, Pratje P, Tajima T, Russon A (2011) Factors affecting reproduction in rehabilitant female orangutans: young age at first birth and short inter-birth interval. Primates 53(2012):181–192

Lappan S, Malaivijitnond S, Radhakrishna S, Riley EP, Ruppert N (2020) The human–primate interface in the new Normal: challenges and opportunities for primatologists in the COVID-19 era and beyond. Am J Primatol 82(8):e23176

Litchfield CA (2008) Responsible tourism: a conservation tool or conservation threat? In: Stoinski TS, Steklis HD, Mehlman PT (eds) Conservation in the 21st century: gorillas as a case study. Developments in primatology: progress and prospects. Springer, Boston, MA. https://doi.org/10.1007/978-0-387-70721-1_4

Macfie EJ, Williamson EA (2010) Best practice guidelines for great ape tourism. IUCN/SSC Primate Specialist Group, Gland, Switzerland. www.primate-sg.org/best_practice_tourism

Marchal V, Hill C (2009) Primate crop-raiding: a study of local perceptions in four villages in North Sumatra, Indonesia. Primate Conservat 24(1):107–116

Meilida YN, Tuah S (2020) Analysis of ecotourism development strategy in Tanjung Puting Province National Park, Central Kalimantan. KnE Social Sci 4(6):966–978. https://doi.org/10.18502/kss.v4i6.6655

Mihalic T (2016) Sustainable-responsible tourism discourse – towards 'responsustable' tourism. J Clean Product 111(16):461–470

Molyneaux A, Hankinson E, Kaban M, Svensson MS, Cheyne SM, Nijman V (2021) Primate selfies and anthropozoonotic diseases: lack of rule compliance and poor risk perception threatens orangutans. Folia Primatol (Basel). 92(5-6):296–305. https://doi.org/10.1159/000520371. Epub 2021 Oct 25. PMID: 34695831

Muehlenbein M, Wallis J (2014) Considering risks of pathogen transmission associated with primate-based tourism. In: Russon A, Wallis J (eds) Primate tourism: a tool for conservation? Cambridge University Press, Cambridge, pp 278–291. https://doi.org/10.1017/CBO9781139087407.021

Paulsen DI (2009) The behavioral and physiological effects of ecotourism on the Sulawesi black macaques at the Tangkoko nature reserve, North Sulawesi, Indonesia. PhD Dissertation. University of Washington, Seattle, WA

Purwanto E (2016) An anti-encroachment strategy for the tropical rainforest heritage of Sumatra: towards new paradigms. Yogyakarta, Indonesia: Tropenbos International Indonesia and UNESCO, p 126

Rijksen HD, Meijaard E (1997) Our vanishing relative: the status of wild Orangutans at the close of the twentieth century. Kluwer Academic Publishers

Riley CM, Russon AE, Wallis J (2015) Primate tourism: a tool for conservation? Primates 56:375–376. https://doi.org/10.1007/s10329-015-0477-z

Russon AE, Russell CL (2005) Orangutan tourism. In: Caldecott J, Miles L (eds) World atlas of the great apes and their conservation. Univ. California Press, Berkeley, CA, pp 264–265

Russon A, Susilo A (2014) Orangutan tourism and conservation. In: Russon A, Wallis J (eds) Primate tourism: a tool for conservation? Cambridge University Press, Cambridge, pp 76–97. https://doi.org/10.1017/CBO9781139087407.007

Russon A, Wallis J (2014) Primate tourism: a tool for conservation? In: Russon AE, Wallis J (eds) . Cambridge University Press

Santos WJ, Guiraldi LM, Lucheis SB (2020) Should we be concerned about COVID-19 with non-human primates? Am J Primatol:e23158

Yaki S (2020) Ecotourism: Enhancing tourism for the benefits of both people and nature. https://www.selamatkanyaki.ngo/our-work/ecotourism/

Shepherd C, Shepherd L (2017) A Naturalist's guide to the primates of Southeast Asia, East Asia and the Indian Sub-continent. John Beaufoy Publishing. ISBN: 978-1-909612-24-2

Shutt K, Heistermann M, Kasim A, Todd A, Kalousova B, Profosouva I et al (2014) Effects of habituation, research and ecotourism on faecal glucocorticoid metabolites in wild western lowland gorillas: implications for conservation management. Biol Conserv 172(2014):72–79

Solomon JN, Gavin MC, Gore ML (2015) Detecting and understanding non-compliance with conservation rules. Biol Conserv 189:1–4

Supriatna J (2006) Conservation programs for the endangered Javan gibbon (Hylobates moloch). Primate Conservat 2006(21):155–162

Supriatna J, Mootnick A, Andayani N (2010) Javan gibbon (Hylobates moloch): population and conservation December. In book: Indonesian Primates

Supriatna J, Ario A (2015) Primates as flagships for conserving biodiversity and parks in Indonesia: lessons learned from West Java and North Sumatra. Primate Conservat 29(1), 123131.

Supriatna J, Dwiyahreni AA, Winarni N, Mariati S, Margules C (2017) Deforestation of primate habitat on Sumatra and adjacent islands, Indonesia. Primate Conservat 31(1):71–82

Supriatna J, MShekelle M, Fuad HAH, Winarni NL, Dwiyahreni AA, Farid M, Mariati S, Margules C, Prakoso B, Zakaria Z (2020) Deforestation on the Indonesian island of Sulawesi and the loss of primate habitat. Global Ecol Conservat 24:e01205

Susilawati SS, Fauzi A, Kusmana C, Santoso N (2020) Strategy and policy in the management of Sumatran Orangutan (Pongo abelii) conservation tourism on the Lawang Hill in the Langkat district of north Sumatera. J Nat Resour Environ Manag 10(1):1–11

SwaraOwa (2021) 2020 in review: humming optimism amidst adversity. https://swaraowa.org/tag/owa-mentawai/. Accessed 23 Mar 2021.

Weber A, Kalema-Zikusoka G, Stevens NJ (2020) Lack of rule-adherence during mountain gorilla tourism encounters in Bwindi impenetrable National Park, Uganda, places gorillas at risk from human disease. Front Public Health 8:1. https://doi.org/10.3389/fpubh.2020.00001

Whittaker D (2006) A conservation action plan for the mentawai primates. Primate Conservat 2006(20):95–105. https://doi.org/10.1896/0898-6207.20.1.95

WWF (1980) Saving siberut: a conservation masterplan. Bogor, Indonesia. World wildlife fund, Indonesia programme

Chapter 4
Rethinking Tolerance to Tourism: Behavioral Responses by Wild Crested Macaques (*Macaca nigra*) to Tourists

D. A. Bertrand, C. M. Berman, M. Agil, U. Sutiah, and A. Engelhardt

Abstract There is an assumption that apparent tolerance of tourists at long-running primate tourism sites indicates habituation and that as a result primates no longer experience negative consequences of prolonged exposure to visitors. We examined effects of tourist presence on stress-related behavior in three groups of critically endangered, wild crested macaques (*Macaca nigra*) exposed to different intensities of tourism in Tangkoko Nature Reserve, Sulawesi, Indonesia. Group R2 has been exposed to research + intensive tourism for over 3 decades, R1 to research + less intensive tourism (1 decade), and PB1 to research only. Almost 740 h of data were collected from 33 adults via focal animal, all occurrence, and 1/0 sampling. All data were analyzed with general linear mixed models. Behavior appeared to be inhibited when tourists were in the forest, but not within groups; all groups vocalized less, exhibited fewer sexual behaviors and displayed fewer self-directed behaviors in months with greater numbers of tourists. When tourists were present vs. absent within groups, females displayed less affiliation, and males and females displayed more aggression, consistent with responses to uncertainty in the presence of tourists. Our results indicate that crested macaque groups exposed to tourism even for decades may not fully habituate to tourists. We tentatively suggest that their behav-

D. A. Bertrand (✉)
Department of Anthropology, University at Buffalo, Buffalo, NY, USA

Georgetown, TX, USA
e-mail: dabertra@buffalo.edu

C. M. Berman
Department of Anthropology, University at Buffalo, Buffalo, NY, USA

Department of Environment and Sustainability, Evolution, Ecology, & Behavior Program, University at Buffalo, Buffalo, NY, USA

M. Agil · U. Sutiah
School of Natural Sciences and Psychology, Liverpool John Moores University, Liverpool, UK

A. Engelhardt
Faculty of Veterinary Medicine, Bogor Agricultural University, Bogor, Indonesia

© The Author(s), under exclusive license to Springer Nature Switzerland AG 2022
S. L. Gursky et al. (eds.), *Ecotourism and Indonesia's Primates*, Developments in Primatology: Progress and Prospects, https://doi.org/10.1007/978-3-031-14919-1_4

45

ioral responses to tourists resemble typical responses of primates to perceived predators posing varying degrees of risk.

Keywords Stress-related behavior · Wildlife tourism · Primates · M. nigra · Tolerance · Aggression · Self-directed behaviors · Predator avoidance

4.1 Introduction

Many wildlife tourism operations that feature nonhuman primates aim to conserve ecosystems, financially benefit local populations, and educate both local people and visitors. Some ecotourism sites have been in operation for many years, often decades. Their target primate species appear to be habituated to tourists, influencing managers, and, at some sites, researchers as well, to assume that the local primate species no longer experience any negative consequences of prolonged exposure to visitors. However, this assumption of habituation is not necessarily accurate, and serious anthropogenic stressors sometimes accompany ecotourism.

4.1.1 Stress and Primate Tourism

Stress is an adaptive response to a perceived threat to survival, i.e., a stressor (Moberg 2000). However, prolonged or frequent exposure to stressors can be maladaptive due to the harmful physiological effects of prolonged or frequent exposure to certain hormones (e.g., glucocorticoids) that are released in response to stressors (Munck et al. 1984; Sapolsky 1992). Thus, uncovering specific prolonged or frequent sources of stress in ecotourist locations is vital. Of important note, glucocorticoids are metabolic hormones which do not function primarily as stress hormones (Beehner and Bergman 2017). Regardless, they are often elevated during exposure to stressors and are widely used as a proxy for the strength of an organism's response to stressors. Measuring potential physiological stress responses under varying conditions is one way to uncover sources of anthropogenic stress in ecotourist sites.

While physiological stress can be measured through various bodily substrates, some sources can be challenging to collect in wild and semi-wild habitats (Sheriff et al. 2011). Thankfully, researchers have connected increases in physiological stress responses to changes in various behaviors (Maestripieri et al. 1992). However, some researchers have reservations about these connections (MacDougall-Shackleton et al. 2019—see details below). Due to this, many ecotourism-focused researchers attempt to measure either physiological stress, potential behavioral stress, or both. For example, Tibetan macaques (*Macaca thibetana*) at Mt. Huangshan, China, display behaviors that may be related to stress more when tourists are present and in response to certain tourist behaviors (Matheson et al. 2006). In addition, howler monkeys (*Alouatta palliata*) in Belize show signs of physiological stress as numbers of tourists increase (Aguilar-Melo et al. 2013). Also, Barbary macaques (*Macaca sylvanus*) scratch more when exposed to large groups of tourists and have higher fecal glucocorticoids after aggressive tourist/macaque interactions (Maréchal et al. 2011). In the present study, we aim to test the general hypothesis that levels of potential stress-related behaviors in three groups of wild, habituated Sulawesi crested macaques (*Macaca nigra; yaki in the local Manadonese dialect*) in Tangkoko Nature Reserve (TNR), NE Sulawesi, Indonesia, are related to aspects of tourism.

4.1.2 Habituation of Primates for Tourism

The local citizens of Batu Putih, a village on the edge of TNR, live alongside multiple *Macaca nigra* social groups. When tourism spread across the globe, the economic value of wildlife became clear. Tourism inside TNR was first formally documented in 1978. Despite exposure to tourists for nearly two decades, Kinnaird and O'Brien found in 1996 that small groups of unknown humans caused *M. nigra* to flee, a potential behavioral response to stress. In order to benefit from tourists flocking to the North Eastern Sulawesi region (Muller 1992; National Resources Management Project 1993), local guides needed macaque groups to stay stationary and on the ground for extended periods of time. The most effective way to achieve this without force was through habituation.

Habituation is a reduction in responses over time as sensory stimuli are perceived as neither adverse nor beneficial (Bejder et al. 2009). It enables an organism to filter out excess environmental stimuli, allowing it to focus on factors critical to its survival. Conversely, stimuli that are perceived as dangerous elicit physiological stress-related responses, which, in excess, can be detrimental to health and/or fitness (Moberg 2000). The ability to recognize situations that pose no threat, and thereby avoid unnecessary physiological stress-related responses, is likely to enhance chances for survival and reproduction. Hence, to reduce possible harm to wild subjects, ensure maintenance of natural behaviors, and facilitate observation, researchers and ecotourism operators seek to habituate target groups (Goodall 1986). In the early days of wild primate observation, a widely used method for "accelerated habituation" was through food provisioning (see Knight 2009; Yamagiwa 2011 for review). Over time, it became clear that provisioning primates as a means for habituation was problematic. Human-directed aggression and crop raiding increased in areas where accelerated habituation was employed (Knight 2009; Yamagiwa 2011). These two behaviors can be detrimental to both nonhuman primates and the people who live around them. Conversely, long-term habituation entails repeated exposure to neutral interactions with observers over time (Tutin and Fernandez 1991; however, see Hanson and Riley 2017, for review of habitation between humans and primates as an intrasubjective process). Researchers often assume that habituation has been achieved when their subjects tolerate their presence, i.e., "the relatively persistent waning of a response as a result of repeated stimulation..." (Hinde 1970).

This definition has led to the assumption that tolerance equals habituation, and habituation equals harmless levels of physiological stress-related responses. Site operators have taken a cue from researchers and use similar methods and criteria to habituate targeted primate groups in an effort to make them more accessible to the growing number of tourists (Johns 1996).

However, habituation is an ongoing behavioral process, requiring considerable long-term scrutiny of both behavioral and biological responses to perceived disturbances (Bejder et al. 2009). As opposed to habituation, what we see more often at tourism sites is apparent "tolerance" of animals to anthropogenic presence (see Blumstein 2016 for review). Indeed, some studies have shown that at several primate field sites, overt primate behavioral changes to human presence decrease quickly (after several months), but less noticeable responses (subtle behaviors or cortisol levels) decrease over a much longer period (Jack et al. 2008; McDougall 2012; Williamson and Feistner 2011). Thus, it may be inaccurate to assume primates are fully habituated and are not experiencing maladaptive levels of physiological stress-related responses. Additionally, many factors are likely to play into the ways in which animals respond to environmental challenges and potential stressors like human disturbance. For example, individuals may respond differently based on their age, sex, dominance status, or personality (Balasubramaniam et al. 2020b; Coleman 2012; Martin and Réale 2008; Sapolsky 2005). Moreover, different characteristics of a potential stressor, such as numbers of tourists or familiar vs. unfamiliar humans, may induce different behavioral responses (Frid and Dill 2002) that may also vary with ecological conditions, including food availability or rainfall (Sheriff et al. 2011). Finally, apparent behavioral tolerance may present in the form of general behavioral inhibition in response to signs of human presence nearby. Such inhibition may not be obvious to observers but may accompany a physiological stress response and/or represent a mild form of threat assessment or avoidance, e.g., vigilance or avoidance of detection. Untangling all these factors is important in any examination of potential stress-related behavior in wild primate groups.

4.1.3 Primate Stress-Related Behaviors

Researchers have identified a number of specific behaviors in primates in laboratory studies that correlate with levels of physiological stress indicators, including glucocorticoids (Maestripieri et al. 1992). These behaviors are used as proxies for the strength of an organism's response to stress and will be referred to throughout this chapter as "stress-related behaviors" or SRBs. These behaviors, although indirect indicators, allow researchers an easy, inexpensive, noninvasive means, to detect minute-to-minute stress-related responses in individuals when hormonal analysis is not possible. The two most commonly studied SRBs are displacement activities: self-scratching and self-grooming (Troisi 2002) both of which increase when captive long-tailed macaques (*Macaca fascicularis*) are injected with an anxiogenic drug meant to induce anxiety (Schino et al. 1996). These behaviors have also been shown to be associated with stressful situations in the wild. For example, intragroup aggression has been shown to increase rates of scratching in wild brown lemurs (*Eulemur fulvus*) (Palagi and Norscia 2011). Additionally, a multifactor study of self-directed behaviors in free-ranging Japanese macaques (*Macaca fuscata fuscata*) provided evidence that self-grooming can act as a displacement activity; it increased in the presence of social uncertainty (Duboscq et al. 2016). However, this has not consistently been the case in primate studies and, in some instances, scratching either decreased or showed no change in the presence of a potential stressor (Maréchal et al. 2016; Ulyan et al. 2006). Other explorations of scratching as an SRB in wild primates suggest that some scratching may simply be due to environmental conditions in the wild, such as increased numbers of biting insects (Duboscq et al. 2016) or ambient temperature and humidity (Ventura et al. 2005). Additional explanations have also come to light, focusing instead on the function of scratching as opposed to its cause. Higham et al. (2009) posited that scratching may be a behavioral coping mechanism, helping ameliorate physiological stress responses. While a study by Whitehouse et al. (2017) suggested that scratching (what they qualify as an observable stress behavior) in primates is adaptive because of its presumed ability to reduce escalated aggression, thereby improving social cohesion. Laméris et al. (2020) came to a similar conclusion when exploring scratching as a behavioral contagion in captive Bornean orangutans (*Pongo pygmaeus*).

Due to the conflicting results surrounding displacement activities such as self-grooming and self-scratching, researchers measure additional behaviors that may suggest increased stress, such as increases in aggression, changes to rates of vocalizations, and decreases in both sexual and affiliative behaviors. For example, Clarke et al. (1996) examined the relationships between aggression, immunological, and hormonal responses associated with social change in two groups of captive rhesus monkeys (*Macaca mulatta*). They found a consistent relationship between aggression and physiological stress indicators in the study group. Specifically, both noncontact aggression and cortisol levels increased during the first 24 hours after an introduction of a new member. Also, in a seminal study, Rowell and Hinde (1963) examined *M. mulatta* behavioral response to a potential threat, a human wearing a "scary" mask. They found that the presence of the mask greatly reduced the frequency of calling (a mix of contact and food calls), which were present in all three other conditions: control, food, and familiar human. A more recent study by Pérez-Galicia et al. (2017) found a decrease in vocalizations in the presence of humans in a group of spider monkeys (*Ateles geoffroyi*) maintained at an island in Mexico. In addition, Mitchell et al. (1991) examined the behavior of zoo-housed, golden-bellied mangabeys (*Cercocebus chrysogaster*). When mangabey groups housed in an enclosure experiencing a moderate number of daily visitors were moved into an enclosure experiencing a low number of daily visitors (a switch from a presumably stressful situation to a less stressful one), sexual behaviors, grooming, and play increased. Also, Chamove et al. (1988) found that 15 species of captive primates showed significantly less affiliative behavior in the presence of visitors (presumably a perceived stressor). Wild primates show similar reactions. For example, proboscis monkey (*Nasalis larvatus*) infants significantly decreased their frequency of social behaviors as numbers of tourists increased, a potential behavioral response that may have been related to stress (Leasor and Macgregor 2014). However, Marty et al. (2019) linked an increase in social behaviors with an increase in SRBs in a group of long-tailed macaques (*Macaca fascicularis*) residing at a site with a high level of anthropogenic impact. While not directly linked to human presence, there is the possibility that increasing social behaviors in times of stress can act as a coping mechanism (e.g., social buffering hypothesis). Conversely, other studies suggest that individuals may use affiliative behavior in more complex ways than just an increase or decrease to cope with potentially stressful situations. For example, Wittig et al. (2008) found that when wild female chacma baboons experienced a stressful situation, as potentially indicated by increases in cortisol, those who reduced their grooming network to a few strong relationships, without necessarily changing their overall grooming rates, displayed greater reductions in cortisol levels than those who maintained a more diverse network made up of weaker relationships. Thus, the size of the social network is important and grooming rates do not always correlate directly with stress. Additionally, Balasubramaniam et al. (2020a) found that semi-urban bonnet macaques (*Macaca radiata*) who spend more time monitoring humans decreased their time spent grooming conspecifics. However, affiliative behaviors with short durations, i.e., lip-smacking, showed no change, indicating that context and behavior have a complex relationship.

4.1.4 Confounds in Measuring Stress-Related Behaviors

Given these complications to using a single SRB, we used a variety of presumed indicators to examine possible associations between macaque stress and aspects of tourism. We define them as "presumed indicators" since changes in these behaviors might have other explanations. For example, macaques may be distracted by tourists and not stressed per se. Distraction might be indicated by an inhibition in behavior, including the self-directed behaviors (SDBs) defined as self-scratching and self-grooming. To complicate matters further, many internal and external factors may change the way a primate responds to a stressor. Given this, controlling multiple possible confounding factors is vital when examining possible stress-related responses to tourism. For example, individual rank influences the way a primate responds to a stressor behaviorally (see review Cavigelli and Caruso 2015). Additionally, a recent study by Woods et al. (2019) assessed visitor-directed aggression in zoo-housed Japanese macaques *(Macaca fuscata)* by rank and found that low-ranking individuals displayed more frequent aggression toward visitors. Considering male primates specifically, the particular reproductive season (Fichtel et al. 2007), the number of fertile females present (Engelhardt et al. 2011, unpublished), or the number actively in a consortship (Bergman et al. 2005) could influence levels of aggression, affiliative behaviors, sexual behaviors, and/or cortisol. Also, male dispersal from their natal group into a new group can influence levels of cortisol, which, in turn, may influence SRBs (*Macaca nigra:* Marty et al. 2017a). In female primates, the number of young infants present in the group could influence levels of aggression. In many species, mothers display heightened aggression in defense of their young (see review Hahn-Holbrook et al. 2011). Alternatively, levels of conspecific affiliation may also shift, as seen in wild ring-tailed lemurs (*Lemur catta*), where affiliative behaviors between adult females increase in the presence of young infants (Nakamichi and Koyama 2000). Additionally, individual reproductive state may influence the way females respond behaviorally due to the physiological links between reproductive hormones and cortisol (Weingrill et al. 2003). Food availability is an external factor of concern. Cortisol is generated to metabolize stored energy reserves when food is scarce (Sapolsky et al. 2000). As such, low food availability is sometimes, but not always, associated with higher fecal glucocorticoid levels (Foley et al. 2001; Pride 2005; Sapolsky 1986). For example, Behie et al. (2010) found a complex relationship between overall food availability, fruit consumption, and cortisol levels in two groups of mantled howlers (*Alouatta palliata*). Specifically, when fruit availability was low, cortisol levels increased. The supposition is that when fruit availability is low, monkeys eat less fruit and therefore obtain less sugar. Indeed, prior literature suggests that the lack of fruit (sugar) may be particularly responsible for adaptive increases in cortisol levels as it increases glucose mobilization (Muller and Wrangham 2004). Whether such diet-related changes in cortisol levels are related to changes in SRBs is unclear. Nevertheless, recognizing the importance of untangling as many factors as possible, we collected data on and controlled for all of these factors during statistical analysis.

4.1.5 Tourism inside Tangkoko Nature Reserve

There is an urgent need to understand the factors, both natural and anthropogenic, that contribute to *Macaca nigra* fitness. The International Union for Conservation of Nature (IUCN 2022) lists *M. nigra* as critically endangered and rates their conservation as a high priority. The study site, Tangkoko Nature Reserve (TNR) is a popular ecotourist location and home to the last remaining, viable population of *M. nigra* (see review Danish et al. 2017).

Previous research has examined the influence of tourism on crested macaque behavior inside the park. As mentioned above, Kinnaird and O'Brien (1996) found that exposure of one macaque group to seven or more tourists often produced fleeing. Additionally, two smaller groups that were less exposed to tourism had either a lower or zero tolerance for tourists. However, this occurred before full habituation (enabling daily, year-round researcher visits) of the macaques. Over a decade later, Paulsen (2009) found that, between two consecutive summers, crested macaque aggressive behaviors increased in frequency and escalated more quickly in the presence of tourists.

In the present study, we tested the hypothesis that levels of stress-related behaviors (SRBs) are associated with aspects of tourism in three habituated groups of wild *M. nigra* in TNR named R1, R2, and PB1. These groups represent a natural experiment, each exposed to different intensities of tourism (R2 = frequently, R1 = moderately, and PB1 = rarely/research only). All three groups have had similar exposure to researchers associated with the Macaca Nigra Project (MNP) for about 15 years. Although the number of researchers in each group varied by day (and was recorded daily), it was limited to 6 for R2 and R1 and 4 for PB1. The two tourism groups (R1 and R2) have been exposed to tourists (and accompanying guides) for about three decades. Additionally, while one group rarely encounters tourists, tourist groups are sometimes loud, and large groups of them can be heard from a great distance. Due to this, we asked not only about the possible effects of the presence of tourists within study groups, but also about possible effects of tourists in the reserve when outside and away from the study groups. MNP is careful to limit the number of researchers in each group, but they have no control over the number of tourists or guides. The TNR tour guide rules (unpublished but distributed, 2015) state that no guide can bring more than four tourists. However, DB and team frequently saw one guide with ten or more tourists or two guides with four or fewer tourists. To control for this in our analysis, we kept track of the number of guides and numbers of tourists, in each macaque group per day.

TNR borders Batu Putih gardens and village homes, both of which present enticing food sources. The macaque social groups were also exposed to guarding of crops when they ventured just outside the park boundaries (generally involving TNR personnel making whooping noises, chasing, or setting off fireworks) with the same variation in frequency as exposure to tourists (R2 = frequently, R1 = moderately, and PB1 = rarely). Due to this, we also recorded daily crop guarding events and avoided recording data during and within 30 min of a crop guarding event in an effort to untangle behavioral responses to tourism from behavioral responses to crop guarding.

4.2 Specific Hypotheses and Predictions

4.2.1 Hypothesis 1: Possible Effects of Tourists in the Forest

We predicted that **(H1)** if exposure to tourists influences the display of SRBs even when tourists are not present within the group, then **(P1)** we will find significant differences in SRBs that are related to levels of exposure to tourism. Specifically, **(H1a)** if groups with more tourist exposure experience more stress than groups with less tourist exposure, then **(P1a)** R2 and R1 will display significantly higher rates of aggression and SDBs and significantly lower rates of sociality and vocalizations than PB1. Also, **(H1b)** if temporal variation in tourist presence in the forest is associated with SRBs, then **(P1b)** we will find significantly higher rates of aggression and SDBs and significantly lower rates of sociality and vocalizations in those months with more tourists in the forest. Moreover, **(H1c)** if the monthly numbers of tourists in each group affect individual groups differently, then (**P1c**) there will be a significant interaction effect between group and numbers of tourists per month; i.e., group responses will be related to their levels of exposure to tourists. Alternatively, (**H1d**) if exposure to tourists in the forest inhibits behavior, then (**P1d**) we will find lower rates of all SRBs in months with many tourists.

4.2.2 Hypothesis 2: Possible Effects of Presence Vs Absence of Tourists within Groups

We predicted that (**H2**) if levels of direct exposure to tourists in the group influence the display of SRBs, then (**P2**) we will find significant differences in SRBs related to their direct exposure to tourists. Specifically, (**H2a**) if the presence of tourists in a focal session is stressful, then (**P2a**) groups will display significantly higher rates of aggression and SDBs and significantly lower rates of sociality and vocalizations during focal sessions with tourists as opposed to those in their absence. Additionally, (**H2b**) if tourist presence itself is stressful, (**P2b**) then groups will display significantly higher rates of aggression and SDBs and significantly lower rates of sociality and vocalizations in those focal sessions that have higher numbers of tourists than those sessions with lower numbers. Also, (**H2c**) if regular exposure to tourists is stressful, then (**P2c**) groups will display significantly higher rates of aggression and SDBs and significantly lower rates of sociality and vocalizations on those days that have higher numbers of tourists over the course of the day than on days with lower numbers. And (**H2d**) if groups with more tourist exposure experience more stress than groups with less tourist exposure, then (**P2d**) R2 will display significantly higher rates of aggression and SDBs and significantly lower rates of sociality and vocalizations than R1. Moreover, (**H2e**) if tourist presence during a focal affects individual groups differently, then (**P2e**) there will be a significant interaction effect between group and tourist presence vs. absence; i.e., group responses will be related to their levels of exposure to tourists. Alternatively, (**H2f**) if the presence of tourists inhibits behavior, then (**P2f**) we will find lower rates of all SRBs when tourists are present than absent, in the group with more tourists (R2), and/or on days with many tourists.

4.2.3 Hypothesis 3: Possible Effects of Researchers and Guides

Finally, we predicted that (**H3**) if macaques respond with stress to familiar (as opposed to unfamiliar) humans, then (**P3a**) we will see increases in SRBs when more researchers are present in the group each day than less, and (**P3b**) we will see increases in SRBs when more guides are present in the group each day than less.

4.3 Materials and Methods

4.3.1 Study Site and Species

Tangkoko Nature Reserve is a location of robust megadiversity (Rhee et al. 2004) that once claimed the highest number of endemic species in any protected area on the island of Sulawesi (MacKinnon and MacKinnon 1980), including *M. nigra*. The population of *M. nigra* in this 8867-hectare nature preserve is most likely the only viable and natural remaining population in the wild (Palacios et al. 2011; Riley 2010; Supriatna and Andayani 2008). The most recent survey indicates that one half of the park supports a population of 1951 or 44.9 individuals per km^2 (Palacios et al. 2011). Another recent survey assessed 61.5 individuals/km^2 (Kyes et al. 2012), which comes close to population numbers of 76 individuals/km^2 from almost 30 years ago (Sugardjito et al. 1989). However, the assessment by Kyes et al. (2012) focused only on the tourism (615 ha) and protected areas (3835 ha), representing approximately one-half of the park.

   M. nigra utilizes a variety of habitats including lowland primary forests, areas of cultivation surrounded by primary and secondary forests, actively logged forests, and dense human habitation and agriculture (O'Brien and Kinnaird 1997; Rosenbaum et al. 1998). Crested macaques are diurnal and semi-terrestrial and spend 59% of their day traveling, foraging, and feeding with the remaining time spent resting and socializing (O'Brien and Kinnaird 1997). Their diet consists primarily of fruit, supplemented with other plant parts as well as invertebrate and vertebrate prey. They live in large multi-male, multi-female groups (O'Brien and Kinnaird 1997), and their social organization is female philopatric and female bonded, with males dispersing at sexual maturity and secondarily at intervals throughout adulthood (Duboscq et al. 2013; Marty et al. 2017b; Reed et al. 1997). Coresident adult males are usually not related and primarily use avoidance when interacting, perhaps indicative of tension due to risky reproductive competition (Tyrrell et al. 2020). Females are more egalitarian and utilize connections with higher ranking males to ensure better foraging options and protection from harassment by lower ranking males (Duboscq et al. 2013; Kinnaird and O'Brien 1996; Reed et al. 1997). They are nonseasonal breeders, with a tendency toward birth peaks between January and May (Engelhardt and Perwitasari-Farajallah 2008).

4.3.2 Tourism

TNR is not a new tourism site. Several macaque social groups have been subjected to daily visits from unfamiliar humans (tourists) for over four decades (MacKinnon and MacKinnon 1980). In the 90s, the popularity of TNR as an ecotourist location experienced a surge (Kinnaird and O'Brien 1996) and has continued to gain in popularity annually (Natalia Kandyoh: Tangkoko Ticket Master, personal communication, 2016). Park rules change frequently, but in general, tourists are required to remain with a guide while in Tangkoko, unless they are going to the beach. However, it is important to note that the macaques can and do frequent the beach. Tourist groups range in size from 2 to 25 individuals, are not required to remain on trails (i.e., they can walk up to a group of monkeys), and can remain in the forest for several hours (from dawn until dusk). Both local and international tourists are allowed to camp inside the park on the beach side. In general, during the low tourist (rainy) season, our research indicated that macaque groups are exposed to an average of 2 tourist groups per day, while during the high tourist (dry) season they can be exposed to an average of 7 groups per day, often more than one at a time. Interacting with the macaques is prohibited, and flash photography is discouraged; however, guides rarely enforce these rules. Additionally, when habituation first began with foreign researchers in the late 70's, food (specifically bananas) was the most commonly used tactic in TNR to increase macaque comfort around unfamiliar humans (Petrus Takasaheng & Alfons Wodi: Tangkoko Guides, personal communication, 2016). While discouraged today, feeding by guides and rangers continues unimpeded (Bertrand, D., personal observation). In addition, the macaques have access to tourist food from garbage bins in the park itself.

4.3.3 Field Methods and Subjects

Data collection took place over 14 months (10/2014–01/2016) within groups R2, R1, and PB1, located inside TNR: group R2 (22–23 adults who experienced research, and frequent tourism), group R1 (40–42 adults who experienced research and moderate tourism), and group PB1 (22–23 adults who experienced research only). Due to their ranging patterns, PB1 encountered tourists on rare occasions. However, either MNP researchers would inform tour guides that the group was restricted, and the tourists would continue through the forest, or the tour guides themselves would recognize the group and direct tourists around them.

We collected behavioral data from 33 adult *M. nigra* (age ≥ 7 years, 15 males and 18 females). MNP categorizes (1) individual adult macaque age as young, middle, or old, (2) female rank (using David's Scores) as either low, middle, or high, and (3) male rank numerically (using Elo ratings), with the number 1 representing the highest ranking male. We obtained both rank and age data from MNP in November 2014, before the start of data collection. In order to ensure that the sample of macaques was comparable across groups, we selected six females from each group with corresponding ages and ranks. They were as follows: three young females—one high ranking, one middle ranking, and one low ranking—and three middle-aged females—one high ranking, one middle ranking, and one low ranking. Female *M. nigra* ranks within their respective groups are linear and generally stable over time (Duboscq et al. 2013). Thus, we anticipated no major changes. Males were selected differently. In R1, we chose six males, four categorized as middle aged and two categorized as old. These macaques were spread out evenly across their linear ranks (1, 3, 5, 6, 9, 10). However, it was not possible to match them with specific males in R2 and PB1 as they had less than six males each. Thus, our selected males from these groups were a mix of ages, ranked 1–5 and 1–4 respectively. Male rank is highly asymmetrical and linear; changes were anticipated (Marty et al. 2017a, b). All ranks were verified and corrected when necessary, by myself, before data analysis using Elo ratings for males and David's Scores for females. Behaviors used for this can be found on Table 4.1.

Table 4.1 *Macaca nigra* behaviors analyzed (All behaviors use standard definitions as defined by the Macaca Nigra Project)

Behavior	Event (When necessary, definitions were adapted from instructions to MNP staff and researchers)
Affiliative—within interval sampling	
Body contact	To stand/sit/lie in contact of another individual more than 5 s. The touch can be with any part of the body.
Social groom	To clean the fur/skin of another one. Hairs are brushed and parted using the hands, while particles are picked up by the hands/mouth.
Ventral embrace	To put arms around another individual often with affiliative facial expressions, face to face with potential ventro-ventral contacts—Unidirectional or bidirectional.
Lateral embrace	To stand side by side but facing in opposite directions, each individual drapes an arm over the other's hips or waist, sometimes both partners inspect the other's genitals—Bidirectional.
Hug	To pass one or both hands/arms around the body of another (multiple combination).
Expressive run	To run away and approach again back and forth another individual while doing affiliative facial expressions and/or repeated soft grunts. Often occurs between females with infants or more rarely between two males involved in affiliative interactions (like mounts and genital grasp).
Play	To engage in relaxed and more or less exuberant patterns including loping gait, running, climbing, swinging, rolling, sliding, jumping, walking on hands, bouncing, pirouetting, toppling, uncoordinated moving, stamping, support shaking, handling, dragging, or throwing an object. Any of these patterns may appear in solitary or social play.
Mock bite	To softly bite the body of a social partner in an action of play or copulation. Occurs as well between males engaged in friendly interactions.
Play face	Relaxed open mouth display. To have teeth bared and the mouth open. It signals the performer's desire to play or to interact friendly.
Genital grasp	Between males or male and juvenile—To grab each other's genitals.
Mount	Between same sex—To climb ventro-dorsally upon a standing partner. The mounter may or may not grip the legs of the partner.
Unknown affiliation	Any affiliative interactions (if not seen in detail); includes but not limited to when focal is hugged, when focal is mounted and when focal is genital grasped.
Aggressive—continuous sampling (* indicates behavior used in rank determination)	
Half-open mouth	To slightly open the mouth with corners drawn back, the lower lip may be retracted and the teeth are partly visible. This display is accompanied by staring. Could also be accompanied by a rattle.

(continued)

Table 4.1 (continued)

Behavior	Event (When necessary, definitions were adapted from instructions to MNP staff and researchers)
Open mouth bared teeth	To open the mouth more or less widely. The teeth and the gums could be exposed. It is accompanied by a threat or a scream. It is used as an attack or in a counterattack.
Jaw movement	To thrust head forward, and the lower jaw is moved up and down rapidly and rhythmically. Could be accompanied by scalp retraction and/or ears flatten and/or a low threat vocalization.
Stare	To look intensively to another individual. Could be accompanied by the half-open mouth. Acts as a mild threat.
*Chase	To run after another individual on more than 2 m to make it run away and/or bite/hit/grab it.
*Bite	To bite another individual.
*Hit	To hit another individual with any limb
*Grab	To catch and hold in a hand a bunch of fur of another one to retain it.
*Push	To push with hand or body to make an individual move away.
Support shake	To stand on a branch or whatever and shake or jump on this support. It is a "show off" behavior.
*Lunge	To make a short run (< 2 m) or a jump toward an individual (warning that could lead to an aggressive behavior.)
Stamp	To make a short run or a jump and finally stand stiff on its forelimb. May be follow by yawning/staring or the half-open mouth display (context of tension or play or aggression to attract attention).
Harassment	Several individuals threat, chase, bite, hit, and grab together against another one. Often during intergroup encounter, one individual of one group is harassed by many of the other. It is also the case when an individual comes to disturb friendly or aggressively a copulative pair.
*Scream	Noisy scream vocalization.
*Flight	To run away in response to another's approach/aggression.
*Crouch	To press body on the ground, 4 limbs flexed in response to another's approach/aggression.
*Protection seeking	An individual threatened or attacked by another approaches and contacts a third individual.
Enlisting	To look several times at a particular individual close by or around to call for support while involved in an aggressive interaction.
Unknown aggression	Any aggressive interaction (if not seen in detail).
Ignore	To stay without reacting when an individual approaches or directs affiliative/aggressive behaviors.
Glance/look away	To avoid making eye contact when an individual interacts or attempts to.
Sexual—continuous sampling	
Sexual present	A female raises or orients the hindquarters toward a male at proximity and may turn the head toward it. Could be accompanied by looking at the male or grasping his genitals. It signals the female's motivation to mate.
Sexual parade	A female presents toward a male several times, passing repeatedly in front of him.
Sexual mount	A male climbs ventrodorsally upon a standing female. He may or may not grip the legs of the female.

(continued)

Table 4.1 (continued)

Behavior	Event (When necessary, definitions were adapted from instructions to MNP staff and researchers)
Mate	A male mounts and introduces his penis in the female's vagina and thrusts. Record when both the male and female are focal.
Silent bared-teeth jaw movement	To vertically retract lips, exposing the teeth, and the lower jaw is moved rhythmically and silently. It is typical of a male when a female approaches or passes by or presents toward the male.
Reaching back	A female grasps the fur, leg, arm, face, or genital of the mating male
Ejaculation	To pulse the anus during mating. Usually occurs only after a series of matings.
Vocalization—continuous sampling	
Contact/lost calls	Louder when far away and/or losing visual contact with the group. If two or more calls are emitted within 5 s of each other, only one "vs" should be recorded.
Alarm call	Short and loud, repeated. Signals a danger (python, people).
Female copulation call	(Repetitive) Call given after copulation, can be loud or low.
Male copulation call	Shrieking vocalization by the male during copulation. Often two parts, but record even if only one part is heard.
Loud call	Adult male vocalization occurs in various contexts (e.g., aggression and mating).
Self-directed behavior—continuous sampling	
Self-groom	To clean its one fur.
Self-scratch	To rake the skin repeatedly using fingers of hands or feet.

In summary, DB and team followed 6 males and 6 females from R1, 5 males and 6 females from R2, and 4 males and 6 females from PB1. There were only two male migrations during the data collection period: one subadult male from R2 into R1 in September 2015 and one unknown subadult male into R1 in mid-November 2015. Because migrants were all subadults, they were not added as focal subjects. We lost one male from PB1 early into the study period (only collected 8 min from him). There was no male to replace him. Additionally, we lost three females at varying time points. From R2, one preselected female died shortly before data collection began. From R1, we lost one female shortly after the study began (we only collected 10 min of data from her) and another female a few weeks later (62 minutes collected from her). For all three, we selected a new female focal of comparable age and rank.

4.3.4 Ethics and Research Permits

The protocols used in the study were approved by the Institutional Animal Care and Use Committee of the University at Buffalo (#ANTO2082N). All protocols adhered to strict ethical standards for wild primate research that were designed in consultation with Macaca Nigra Project and the Institut Pertanian Bogor to comply with the legal requirements of Indonesia. All research and physiological sample collection/

shipment permits were obtained and renewed on the appropriate timelines from Balai Konservasi Sumber Daya Alam (Conservation of Natural Resources in North Sulawesi), Kementerian Riset dan Teknologi (Indonesian Ministry of Research & Technology), and Direktorat Jenderal Konservasi Sumber Daya Alam dan Ekosistem (Indonesian Directorate General of Nature Resources and Ecosystem Conservation).

4.3.5 Behavioral Data Collection

A team of six assistants and DB collected behavioral data related to the macaques' responses to tourist presence and characteristics. The assistants comprised three recent college graduates from the United States, and three recent college graduates from Indonesia. We conducted two-minute focal follows to record rates of behavioral stress indicators (SRBs): self-directed behaviors (SDBs, including self-scratching and self-grooming), aggression, vocalizations, and sexual behaviors. These short focal sessions had proven effective in preliminary research. The macaque groups tended to spread out, with tourists moving in between smaller subgroups. Therefore, tourists may have only been within 10 m of a focal monkey for a few minutes at a time. Additionally, a large portion of R1 and R2's home range was comprised of secondary forest with thick scrub, making longer focal sessions difficult.

We also used 1/0 sampling to record the occurrence of affiliative behaviors within the two-minute focal session (see Table 4.1 for complete list and definitions). Each focal session began with a point time sample to record the presence and absence of tourists as well as tourist characteristics (number of, gender, age, and foreign/domestic distinction). All members of the team participated in interobserver reliability testing for identity recognition and the full ethogram of behaviors. For identity recognition, long-term observers (Research Manager and permanent field assistants) were the standards. No statistical test was used. We were tested until we could identify 100% of macaques in each group. For behavioral testing, DB was the standard and we used Cohen's kappa coefficient as our reliability measure. All team members reached reliability levels of at least Kappa 0.96.

The order of focal sessions was randomly assigned for each day of the week, before the week's observations, using an online randomization generator. If two assistants were in the same group, the focal subjects were split equally between them. Thus, an observer was responsible for between 5 and 12 focal macaques on any given day. Each focal subject had at least 30 min between each of their focal sessions. We achieved this easily because, regardless of how many focal subjects were assigned to an observer; it often took several minutes to find the next focal subject on the list. If a focal subject was not found within 15 min, the observer moved to the next subject down the list. Additionally, if a focal subject was lost before the 1-min 45-s mark of the 2-min focal session and the observer did not find their lost focal subject within 5 min, the focal session was deleted. See Table 4.2 for total focal hours collected per individual and group in each condition.

Table 4.2 Number of focal hours and focal sessions analyzed

Focal hours for Hypothesis 1

Group R1			Group R2			Group PB1		
Macaque ID	Hours	Sessions	Macaque ID	Hours	Sessions	Macaque ID	Hours	Sessions
Ak	17.69	531	An	24.69	741	Aa†	14.07	422
Cu†	15.06	452	Fd†	24.56	737	Ba†	15.05	452
Ej	15.02	451	Id†	24.30	729	Bp†	15.51	465
Fu†	18.28	548	Od†	23.07	692	Cp†	15.08	452
Gs†	16.84	505	Qd†	22.42	672	Fm	15.52	466
Hs†	17.69	531	Rm	20.16	605	Ql	7.55	226
Kn	18.46	554	Rn	24.02	721	Rp†	14.30	429
Ll	17.70	531	Td†	24.24	727	Uk	14.67	440
Mm	18.04	541	Tl	25.18	755	Ul	15.29	459
Nu†	17.21	516	Wj	26.53	796	Up†	16.39	492
Om	16.24	487	Zd†	24.75	743			
Qs†	15.75	473						
Total male	100.83	3025	Total male	120.49	3615	Total male	53.04	1591
Total female	103.14	3094	Total female	143.43	4303	Total female	90.4	2712
Total group	203.97	6119	Total group	263.92	7918	Total group	143.33	4300

Mean ± SE hours per focal subject = 18.53 ± 1.055

Total hours for hypothesis 1 = 611.33

Total hours analyzed = 739.5

† denotes female

Focal hours for Hypothesis 2 and Hypothesis 3

Group R1			Group R2			Group PB1	
Macaque ID	Hours	Sessions	Macaque ID	Hours	Sessions	Macaque ID	Hours and sessions
Ak	20.73	622	An	31.2	937	Aa†	N/A
Cu†	17.12	514	Fd†	30.6	918	Ba†	N/A
Ej	17.19	516	Id†	30.0	900	Bp†	N/A
Fu†	21.22	636	Od†	29.1	873	Cp†	N/A
Gs†	18.92	568	Qd†	27.2	817	Fm	N/A
Hs†	19.56	587	Rm	24.5	734	Ql	N/A
Kn	21.74	652	Rn	31.7	950	Rp†	N/A
Ll	20.91	627	Td†	30.3	910	Uk	N/A
Mm	21.41	642	Tl	34.2	1025	Ul	N/A
Nu†	20.35	610	Wj	34.4	1031	Up†	N/A
Om	18.42	553	Zd†	30.5	916		
Qs†	17.86	536					
Total male	99.66	2990	Total male	155.99	4680	Total male = N/A	
Total female	135.76	4073	Total female	177.7	5331	Total female = N/A	
Total group	235.42	7063	Total group	333.69	10,011	Total group = N/A	

Mean ± SE hours per focal subject = 24.74 ± 1.293

Total hours for hypothesis 2 and hypothesis 3 = 569.11

† denotes female

Table 4.3 Fruit availability index at Tangkoko Nature Reserve 2014–2015

$\sum_1$ = Sum of log food scores	N = # trees measured	Log mean food abundance = $\sum_1/N$
$\sum_2$ = sum of plots	X = # of trees sampled in each species in all plots	Mean density = $\sum_2/X$
FAI = $(\sum_1/N)*(\sum_2/X)$		

Several variables were tabulated after data collection was complete. We calculated the number of tourists in the park each month by summing the number of tourists present in all groups each day. While it is possible that some tourists were "double counted' and that others never visited a group, e.g., beach goers or tarsier tourists, these problems were probably minimal because groups were often far apart from each other and it was unlikely that tourists visited both on the same day. In addition, when the groups were close together, no focal sessions were recorded because it was considered to be an "intergroup encounter," thus potentially biasing the measurement of stress-related responses to tourists. We calculated the number of tourists in each group each day by summing the actual number of tourists present in the group from the moment the macaques came down out of their sleeping tree to the moment they were up in their sleeping tree. We calculated the number of guides each day by summing the numbers of guides present in each group from the moment the macaques came down out of their sleeping tree to the moment they were up in their sleeping tree (Table 4.3).

Finally, as we were concerned about high-energy food sources and their potential effect on physiological stress responses, we calculated a fruit availability index (FAI). We calculated this with phenology data collected by Macaca Nigra Project staff using a method designed by Dr. Oliver Schulke and a formula modified from a food availability index derived by Sari (2013). It included all known fruit species foraged by the monkeys except mango trees (spp *Mangifera*), which were not included in the phenology dataset. Furthermore, the measurement of the productivity of coconut trees (*Cocos nucifera*) was problematic partly because the total number of coconut trees was unavailable. Hence, it was not possible to calculate their density, a critical component in the FAI formula. Alternatively, coconut fruits were removed from the formal FAI calculations and instead marked as either present or absent during a data collection month. Recognizing the importance of coconuts as a high-energy food source, we transformed our FAI to a mean rank measure to include the potential use of coconuts during a given month. The formula used in the present study was as follows: FAI = (Sum of log food scores/# trees measured) / (# of trees sampled in each species in all plots/20 plots) = $(\sum_1/N)*(\sum_2/X)$.

4.3.6 Data Analysis

We tested the predictions of H1–H3 using general linear mixed model (GLMM) analysis from the LME4 package version 1.1–12 in R 3.3.3 [Release Version 1.68 (7328)]. One set of models was run for H1 (Model 1), and another set (Model 2) was run for H2 and H3. Separate models were run for males and females and for each SRB. Our unit of analysis was the individual focal session. Each stress-related behavior was entered as the response variable in a separate model. SRBs analyzed were rates of aggression, rates of self-directed behaviors (SDBs: self-scratching and self-grooming), presence or absence of affiliative behaviors, rates of sexual behaviors, and rates of vocalizations. See Table 4.4 for a list of specific model factors.

Before each GLMM model was built, collinearity among variables was tested to ensure all fixed and random effects were not highly correlated with one another. All fixed and random factors had VIF factors below 3. After this, data were explored with qnorm functions to identify the appropriate distribution for GLMM family selection. All response variables indicated Poisson distributions, except for affiliative behaviors, which were collected using 1/0 sampling, indicating a binomial distribution. All behavioral responses explored with Poisson GLMM models included code to offset by the total time, in seconds, of each focal session, providing true rates of behavior.

For each type of analysis, the first model run was always the null (consisting of the response variable, the control factor of fruit availability, and the random factor of macaque ID). The second model run was the full factor model. When evaluating differences between full models and null models, we applied Bonferroni corrections separately to each set of five models within each dataset (those for H1 for males, H1 for females, H2 for males, and H2 for females) by setting a critical level of $0.05/5 = 0.01$ to each model. All full models were significantly different from the null, indicating that one or more fixed effects in the full model were associated with variation in the response factor. Otherwise, criteria for significance were $p \leq 0.05$. All models were checked for overdispersion by testing if the model deviation was larger than the mean. If a model was significantly overdispersed, this indicated that there was greater variability (statistical dispersion) in a dataset than would be expected based on a given Poisson statistical. In these cases, we corrected overdispersion by creating an additional random intercept for each focal session (Elston et al. 2001).

When running GLMM models on behavioral data that fit a Poisson model, it was not always possible to retain this Poisson family due to either a lack of convergence or eigenvalue errors. Convergence errors indicate that the model has too many factors for its sample size and cannot be fit. Eigenvalue errors indicate that one of the variables has a range that is skewed far from the response variable range. In order to correct for convergence errors, we increased GLMM model iterations. If convergence errors did not disappear with the third iteration increase, the response variable was collapsed into a binomial form. In order to correct for eigenvalue errors, it was necessary to rescale one or more continuous variables to a smaller range by converting the data to standard scores and then rescaling to a specified mean. If eigenvalue errors did not disappear, the model would not proceed and we collapsed the response variable data into a binomial form.

Table 4.4 Description of model factors

Factor name	Definition	Present in model:
Group	Social group (R1, R2, or PB1)	1 and 2
Rank/Elo	Dominance rank of focal subject	1 and 2
Monthly_Tourist	Sum of the number of tourists present each month in the forest	1 and 2
fai_rank	Fruit availability index transformed into a ranking system	1 and 2
T-Den	Number of tourists present during focal session	2
Type	Type of focal session, (tourists present vs. free from any anthropogenic condition other than researchers	2
Num_of_Tourists	Daily number of tourists in the focal subject's group	2
Number_of_guides	Daily number of guides present in the focal subject's group	2
Num_of_Researchers	Daily number of researchers present in the focal subject's group	1 and 2
Repro_state.Coll	Individual female macaque reproductive state	1 and 2
Sum.CG.Events	Sum of numbers of days in a month that had one or more crop guarding event	1 and 2
CropGuard	Occurrence of daily crop guarding in the focal subject's group (yes/no)	1 and 2
Fertile	Number of fertile females present each month	1 and 2
Infant	Number of young infants present each month	1 and 2
Monkey[a]	Focal subject ID	1 and 2
Helper	Random intercept to control for over dispersion, if present	1 and 2
C.Con_Agg	Rate of conspecific-directed aggression	1 and 2
Vocal	Rate of vocalizations	1 and 2
NewAffiliative	Occurrence of affiliative behavior in a focal session (yes/no)	1 and 2
SDB.Stress.Binom	Occurrence of stress behaviors in a focal session (yes/no)	1 and 2
SDB.Stress	Rate of stress behaviors	1 and 2
Sexual_Beh	Rate of sexual behaviors	1 and 2

[a]Each model included the focal's ID as a random factor

4.4 Results

4.4.1 Possible Effects of Tourists in the Forest

Results for H1, which involved examining behavior when no tourists were present in the group (but were present in the forest), are shown in Table 4.5. In **Prediction 1a,** we asked if PB1, as the research only group, displayed lower rates of SRBs than R1 and R2. Consistent with this prediction, PB1 females scratched less than R2 females (Z = −1.99, p < 0.0463). However, PB1 males scratched more than R1 males (Z = 3.76, p = 0.046), and PB1 females affiliated less than R2 females (Z = − 2.48, p = 0.013), a finding contrary to our prediction.

In **Prediction1b,** we asked whether the groups may have been affected by the presence of varying numbers of tourists in different months in the park. We found the following main effects: Males vocalized less (Z = −1.98, p = 0.047), showed fewer SDBs (Z = −5.91, p < 0.001), and displayed fewer sexual behaviors (Z = −2.68, p = 0.007) in months with greater numbers of tourists. In addition, females aggressed less (Z = −2.53, p = 0.011), vocalized less (Z = −4.64, p < 0.001), showed fewer SDBs (Z = −4.42, p < 0.001), and displayed fewer sexual behaviors (Z = −2.95, p = 0.003) in months with greater numbers of tourists. These results did not support the prediction that individuals would display more stress-related behaviors in months with more tourists (**P1b**), but they were consistent with the prediction that (**P1d**) greater numbers of tourists may inhibit macaque behavior.

Finally, we asked (**P1c**) whether responses to numbers of tourists per month varied by group or whether all groups responded in a similar manner. Given main effects suggestive of general inhibition of SRBs, we predicted that PB1 would have responded more slowly or less intensely to numbers of tourists each month than R1 and R2. The results for vocalizations were consistent with these predictions as indicated by significant interaction effects for group by monthly tourist numbers. As monthly numbers of tourists increased, R2 males decreased their rate of vocalization faster than either PB1 or R1 (R2 vs. PB1: Z = −2.86, p = 0.004; R2 vs. R1: Z = −2.77, p = 0.006), R1 and R2 females decreased rates of vocalizations faster than PB1 (R1 vs. PB1: Z = −2.28, p = 0.022; R2 vs PB1: Z = −4.40, p < 0.001), and R2 females decreased rates of vocalizations faster than R1 (Z = −1.93, p = 0.049). However, results for other measures were generally in the opposite direction from predicted. PB1 females decreased rates of aggression faster than R1 (Z = −2.51, p = 0.012); PB1 females decreased rates of sexual behaviors faster than either R1 or R2 (PB1 vs. R1:Z = −2.11, p = 0.035; PB1 vs. R2: Z = −1.96, p = 0.048); PB1 males decreased their rates of SDBs faster than either R1 or R2 (PB1 vs R1:Z = −4.45, p < 0.001; PB1 vs. R2: Z = −3.23, p = 0.001), and R2 decreased their rates of SDBs faster than R1 (Z = −2.13, p = 0.033); PB1 and R2 females decreased rates of SDBs faster than R1 females (PB1 vs. R1:Z = −4.27, p < 0.001; R2 vs. R1: Z = −3.45, p < 0.001). R1 females were the only ones to increase their scratching rate.

Table 4.5 Model 1—responses to tourists in the forest

Fixed effects	Dependent Variables									
	Aggression		Vocalizations		Affiliative		SDBs		Sexual	
	SD	z	SD	z	SD	z	SD	z	SD	z
Males										
# Tourists/mo	0.01	0.65	0.01	−1.98*	0.01	−0.75	0.01	−5.91***	0.02	−2.68**
Group	Chisq = 3.07		Chisq = 2.68		Chisq = 2.74		Chisq = 13.49**		Chisq = 1.10	
PE1 vs R1	0.24	−1.57	0.23	−1.21	0.22	1.68	0.08	3.76***	0.49	1.06
PE1 vs R2	0.24	−1.66	0.24	−1.62	0.23	0.71	0.08	0.33	0.50	0.79
R1 vs R2	0.21	−0.14	0.23	−0.51	0.20	−1.05	0.07	−3.74**	0.45	−0.27
Group x no. tourists/mo	Chisq = 0.62		Chisq = 11.93**		Chisq = 1.97		Chisq = 19.82***		Chisq = 3.66	
PB1 vs R1	0.01	0.76	0.02	0.88	0.01	−1.12	0.01	−4.45***	0.02	−1.67
PB1 vs R2	0.01	0.52	0.01	2.86**	0.01	−0.22	0.01	−3.23**	0.02	−0.71
R1 vs R2	0.01	−0.01	0.01	2.77**	0.01	1.26	0.01	2.13*	0.02	1.51
No. researchers	0.02	−1.58	0.02	−1.60	0.02	0.39	0.01	0.87	0.03	0.50
Rank	0.05	0.85	0.05	1.19	0.06	0.85	0.02	−0.45	0.09	−1.33
Fertile females	0.03	3.35***	0.03	0.66	0.04	−1.46	0.01	−2.02**	0.05	1.19
# crop guard events/mo	0.00	−0.58	0.00	2.28*	0.00	−1.33	0.01	−5.76***	0.01	4.03***
Fruit availability index rank	0.02	−1.20	0.02	4.13***	0.02	1.86	0.01	13.87***	0.03	7.36***
Females										
# Tourists/mo	0.01	−2.53*	0.01	−4.64***	0.01	−1.15	0.01	−4.42***	0.04	−2.95**
Group	Chisq = 1.56		Chisq = 3.65		Chisq = 6.19*		Chisq = 4.26		Chisq = 2.34	
PB1 vs R1	0.20	−0.05	0.21	−1.21	0.13	−0.90	0.08	−0.57	0.76	1.54
PB1 vs R2	0.20	−1.03	0.23	−1.83	0.14	−2.48*	0.09	−1.99*	0.79	1.30
R1 vs R2	0.18	−1.13	0.20	−0.75	0.12	−1.84	0.08	−1.66	0.71	0.22
Group x no. tourists/mo	Chisq = 6.95*		Chisq = 49.66***		Chisq = 1.02		Chisq = 20.42***		Chisq = 5.00	
PB1 vs R1	0.01	−2.51*	0.01	2.28**	0.02	−0.85	0.01	−4.27***	0.04	−2.11*
PB1 vs R2	0.01	1−.20	0.01	4.43***	0.02	−0.09	0.01	−1.60	0.04	−1.96*
R1 vs R2	0.01	1.86	0.01	1.93*	0.02	0.92	0.01	3.45***	0.03	0.69

(continued)

Table 4.5 (continued)

Fixed effects	Dependent Variables									
	Aggression		Vocalizations		Affiliative		SDBs		Sexual	
Males	SD	z	SD	z	SD	z	SD	z	SD	z
No. researchers	0.02	1.93	0.02	1.27	0.02	−1.76	0.01	**−1.98***	0.05	0.68
Rank	Chisq 3.28		Chisq = 1.00		**Chisq = 6.23***		Chisq = 3.64		Chisq = 2.63	
High vs middle	0.15	−1.88	0.12	−0.98	0.09	1.37	0.09	−1.92	0.61	0.87
High vs low	0.20	−1.19	0.17	−0.27	0.12	−1.43	0.07	−0.42	0.82	1.76
Middle vs low	0.18	−0.19	0.15	−0.49	0.12	**2.66****	0.08	1.76	0.77	−1.18
Reproductive state	**Chisq = 70.31***		**Chisq = 18.98***		Chisq = 1.18		Chisq = 1.46		**Chisq = 226.01***	
Cycling vs lactating	0.08	**−4.99***	0.07	−1.96	0.06	0.24	0.02	−0.11	0.44	**9.09***
Cycling vs pregnant	0.09	**2.09***	0.06	**−2.67****	0.06	−0.70	0.03	−1.18	0.17	**7.06***
Pregnant vs lactating	0.07	**−8.14***	0.05	0.70	0.06	1.05	0.03	0.88	0.45	**6.01***
# of infants	0.02	−0.02	0.02	**−1.99***	0.01	−1.80	0.01	**−8.33***	0.05	1.07
# crop guard events/mo	0.00	−0.47	0.00	**5.55***	0.01	−1.38	0.01	**−3.94***	0.01	−1.46
Fruit availability index rank	0.02	1.53	0.01	0.34	0.01	**4.36***	0.01	**11.71***	0.05	**3.71***

*** p ≤0.001, ** p ≤0.01, * p ≤0.05

4.4.2 Presence Vs Absence of Tourists

Results for **Hypothesis 2**, which involved examining behavior of the two tourism groups, are shown in Table 4.6. In **Prediction 2a,** we asked if tourist presence vs. absence in a focal session influenced the display of SRBs. We found that that females displayed less affiliation (Z = −2.15, p = 0.031) and more aggression (Z = 3.33, p = 0.001) during focal sessions in which tourists were present as opposed to absent. In addition, males aggressed more (Z = 2.60, p = 0.009). These results support the prediction that macaques responded to the presence of tourists during a focal session with increases in some stress-related behavior (**P2a**) but not with general inhibition of behavior (**P2f**). There were no other measures with significant differences. In (**P2b**), we examined whether the number of tourists present in each focal session was related to the display of SRBs. Only females displayed any variation in SRBs with numbers of tourists within the group. When more tourists were present, they had higher rates of SDBs (Z = 3.75, p < 0.001). In **Prediction 2c,** we asked if the total numbers of tourists present in the group each day were related to the display of SRBs. Both males and females displayed significantly lower rates of SDBs when the number of tourists present each day were higher (males: Z = −3.01, p = 0.003; females: Z = −4.69, p < 0.001). Additionally, females displayed significantly lower rates of vocalizations when the numbers of tourists present each day were higher (Z = −2.86, p = 0.004). Although these results were consistent with the idea that tourists within groups inhibited scratching and vocalizing, results for no other SRBs reached statistical significance. As such, these results are not consistent with predictions for stress (**P2c**) and represent weak evidence of inhibition (**P2f**). In (**P2d**), we predicted that the group that experienced more direct exposure to tourists would display more behavioral stress indicators. However, we found no significant differences in SRBs between the two groups. Finally, **P2e** asked whether the two groups differed in their responses to the presence vs absence of tourists or whether both groups responded in a similar manner. Although R2 was exposed to tourists more frequently than R1, there were no significant interactions between tourist presence vs. absence and group for any SRB.

4.4.3 Researchers and Guides

Finally, results for **Hypothesis 3** (using data from Model 2) in which we asked whether the daily numbers of researchers (**P3a**) and guides (**P3b**) present in each group were related to the display of SRBs are shown in Table 4.6. We found that males displayed less aggression when a greater number of researchers were present (Z = −2.24, p = 0.025). However, females displayed higher rates of aggression (Z = 2.91, p = 0.001) and lower rates of SDBs (Z = −3.55, p < 0.001) when a greater number of researchers were present. No other measures varied significantly with numbers of researchers. As such, these results do not represent strong evidence for

Table 4.6 Model 2 and Model 3—responses to tourists in social groups R1 and R2

Fixed effects	Dependent variables									
	Aggression		Vocalizations		Affiliative		SDBs		Sexual	
Males	SD	z	SD	z	SD	z	SD	z	SD	z
Tourist yes/no during focal	0.14	2.60**	0.19	−0.68	0.19	0.48	0.06	−0.23	0.22	0.88
# Tourists/mo	0.01	−0.10	0.01	−3.73***	0.01	−1.20	0.00	−10.40***	0.02	−1.48
# Tourists/day	0.03	0.93	0.03	1.12	0.01	−0.83	0.01	−3.01**	0.46	−0.27
# Of tourists in each focal	0.01	0.25	0.02	−1.13	0.01	−0.04	0.00	1.18	0.03	−1.22
Group R1 vs group R2	0.23	0.54	0.22	0.97	0.20	−0.34	0.15	−1.29	0.39	0.81
Group x tourist yes/no	Chisq = 1.57		Chisq = 0.80		Chisq = 0.14		Chisq = 0.57		Chisq = 0.10	
R1 vs R2	0.13	1.25	0.18	0.89	0.17	−0.37	0.06	−0.78	0.22	0.31
# Of guides	0.07	−1.20	0.10	0.21	0.10	−2.00*	0.03	−2.24*	0.11	1.91
No. researchers	0.03	−2.24*	0.03	−1.63	0.03	1.46	0.01	0.69	0.04	0.73
Rank	0.07	−0.66	0.07	1.65	0.07	0.92	0.02	0.64	0.12	0.34
Fertile females	0.04	1.53	0.04	1.20	0.05	−1.41	0.01	−1.02	0.06	0.04
Crop guard yes/no during day	0.07	1.52	0.08	2.66**	0.08	0.57	0.03	3.59*	0.12	1.44
Fruit availability index rank	0.02	0.23	0.02	2.93**	0.02	1.58	0.01	11.56***	0.04	7.62***
Females	**SD**	**z**	**SD**	**z**	**SD**	**z**	**SD**	**z**	**SD**	**z**
Tourist yes/no during focal	0.18	3.33***	0.17	0.77	0.16	−2.15*	0.07	−1.85	0.55	0.86
# Tourists/mo	0.01	−2.44*	0.01	−4.54***	0.01	−2.90*	0.00	−8.79***	0.03	−3.62**
# Tourists/day	0.03	1.61	0.03	−2.86**	0.01	−0.69	0.01	−4.69**	0.09	−1.60
# Of tourists in each focal	0.01	0.68	0.01	0.45	0.01	−1.13	0.00	3.75***	0.03	−0.16
Group R1 vs group R2	0.17	−1.14	0.21	0.67	0.11	−1.89	0.10	−1.47	0.73	−0.10
Group x tourist yes/no	Chisq =0.52		Chisq = 2.35		Chisq = 3.19		Chisq = 0.01		Chisq = 0.01	
R1 vs R2	0.16	0.72	0.15	1.32	0.14	−1.78	0.07	−0.11	0.46	0.05
# Of guides	0.11	−3.01**	0.09	−0.83	0.08	−0.38	0.04	0.08	0.30	−0.93

No. researchers	0.03	2.91**	0.03	1.38	0.02	−0.07	0.01	−3.55***	0.07	0.57
Rank	**Chisq = 8.02***		Chisq = 1.44		Chisq = 3.20		Chisq = 1.05		Chisq = 2.77	
High vs middle	0.30	−1.63	0.40	0.31	0.20	−1.22	0.14	−1.03	1.62	1.69
High vs low	0.16	−3.25**	0.21	−1.10	0.10	1.20	0.07	−0.43	0.72	0.15
Middle vs low	0.29	−0.92	0.38	−0.93	0.19	1.83	0.13	0.83	1.55	−1.69
Reproductive state	**Chisq = 55.12***		**Chisq = 40.56***		Chisq = 0.49		Chisq = 5.15		**Chisq = 84.08***	
Cycling vs lactating	0.12	−4.52***	0.12	−0.63	0.09	−0.55	0.03	0.99	0.76	5.56***
Cycling vs pregnant	0.12	1.61	0.11	−3.15**	0.09	−0.69	0.03	−1.73	0.22	5.77***
Pregnant vs lactating	0.10	−2.73***	0.10	2.67**	0.08	0.12	0.04	2.18*	0.75	3.92***
# Of infants	0.02	−1.67	0.02	2.28*	0.02	−0.30	0.01	−3.65***	0.06	1.55
Crop guard yes/no during day	0.08	−2.14*	0.07	−0.50	0.06	−2.28*	0.03	3.21**	0.18	−2.73***
Fruit availability index rank	0.02	1.19	0.02	1.74	0.02	3.13**	0.01	8.77***	0.07	2.97**

*** p ≦ 0.001, ** p ≦ 0.01, * p ≦ 0.05

either increases in stress-related behavior or behavioral inhibition. In contrast, we found limited evidence of behavioral inhibition related to numbers of guides: Males displayed affiliative behaviors in fewer focal sessions ($Z = -2.00$, $p = 0.045$) and lower rates of SDBs ($Z = -2.24$, $p = 0.025$) when a greater number of guides were present. Females displayed lower rates of aggression when a greater number of guides were present ($Z = -3.02$, $p = 0.002$).

4.5   Discussion

This study aimed to test the general hypothesis that levels of stress-related behaviors in groups of wild *M. nigra* in Tangkoko Nature Reserve (TNR), NE Sulawesi, Indonesia, are related to aspects of tourism. We collected data from three habituated groups with varying levels of exposure to tourism. Overall, our results suggest that wild crested macaques are behaviorally inhibited when more tourists are present in the forest but not present within groups. In addition, they show signs of both inhibition and increases in stress-related behaviors when tourists are present directly in social groups. We tentatively suggest that these responses can be viewed within the framework of typical responses of primates to perceived predators posing varying degrees of risk. Below, we develop this argument in greater detail.

In those months where greater numbers of tourists were present in the forest, we saw, in general, an inhibition of macaque behaviors: Males vocalized less and displayed fewer sexual behaviors and SDBs, and females vocalized less, aggressed less, and showed fewer SDBs. In addition, several measures suggested that degrees of inhibition in the three groups were associated with their levels of direct exposure to tourists. PB1, the group that was exposed to tourists the least, appeared to react more strongly than the other two groups to increased numbers of tourists in the forest each month; PB1 generally showed more intense decreases in aggression, sexual behavior, and SDBs than the other groups. This raises the hypothesis that PB1's relative lack of direct exposure to tourists may have led to more intense behavioral inhibition to their presence in the forest. Vocalizations, however, showed the opposite association with PB1 decreasing its vocalization rates less intensely in response to numbers of tourist in the forest. As such, this finding and those for differences between the two tourist groups complicate this interpretation. Both males and females in the more highly exposed R2 group showed more intense decreases in vocalizations than those in R1. Of note, while our vocalization measures analyzed here included contact calls, long calls, and sexual calls, contact and long calls made about 96% of the total vocalizations.

In contrast, when examining behavior when tourists were present vs. absent within groups during focal sessions, we found some behavioral differences consistent with the idea that direct exposure to tourists is associated with immediate increases in stress-related behavior. Both sexes displayed significantly higher rates of aggression toward conspecifics, and females displayed significantly lower rates of sociality when tourists were present within the group. In addition, when more tourists were present, females displayed higher rates of SDBs. At the same time, other results suggested some evidence of inhibition of behavior when the total numbers of tourists present in the group each day were high; both males and females displayed lower rates of SDBs and females vocalized less, raising the possibility that large numbers tourists over the course of a day may moderate responses somewhat to tourists within groups. This possibility could be tested in the future by looking at changes in responses to tourist groups on a given day as numbers of tourists accumulate over the course of the day. It may also be useful to look at the timing of tourist visits as well as their numbers. On some days, the groups of tourists waited for the macaques at their sleeping trees until they awoke, while on other days the macaques would not encounter any tourists until late afternoon. These changes in visiting tourist patterns not only introduce uncertainty, but may also alter baseline tolerance levels.

Evidence of both behavioral inhibition and increases in typical stress-related behaviors such as increased aggression and SDBs requires a careful examination of the context surrounding each type of response. Inhibition of behavior in one context with an increase in aggression in another may seem surprising, but may be possible to interpret within a framework of responsiveness to different levels of perceived risk to predators, as described by Roelofs (2017). Roelofs posited that as predator threat levels increase, animals move from freezing to fight-or-flight responses. As such, we tentatively suggest that unfamiliar humans trigger mild predator avoidance responses in wild crested macaques and further that they may respond with different behaviors to different levels of perceived risk. These responses appear to have three stages: inhibition, increased SDBs, and increased aggression toward/flight from perceived predators.

While large hawks and pythons are known predators of this population, the macaques' top predator is currently humans through poaching and timber harvesting (Hilser et al. 2013; Supriatna et al. 2020). When the threat of a poacher becomes immediate, macaques typically alarm call and flee into high trees (Diswal Takasaheng: Tangkoko Guide, personal communication, 2015). Due to their experience with poachers, wild crested macaques in Tangkoko may also view unfamiliar tourists as threatening to some extent. Unfamiliar tourists resemble poachers in some respects but not others. Whereas researchers and guides visit groups frequently (with researchers wearing distinctive shirts), poachers and unfamiliar tourists visit rarely and do not wear distinctive clothing. Additionally, only poachers typically bring dogs. Thus, unfamiliar humans likely represent the unknown and add uncertainty to the context.

While there is still much to be learned about predator avoidance/defense in primates, some similarities in predator avoidance behaviors are found across primate species, including vigilance (Stanford 2002). Vigilance is generally defined as a visual scanning of the area (Beauchamp 2015), but it is also typically accompanied by a "freezing," or a general inhibition of behavior (Roelofs 2017). Moreover, behaviors considered to indicate "anxiety," such as scratching, tend to decrease in the presence of behaviors considered to indicate "fear" such as freezing (Barros et al. 2004) As predator presence becomes more evident or proximate, primates may shift to a 2nd stage of predator defense including clumping of individuals (e.g., females gathering infants and moving closer to males), alarm calling (Stanford 2002), and in some cases may increase rates of self-scratching (Palagi and Norscia 2011). However, scratching does not follow this pattern in all primate species. Neal and Caine (2016) found that captive common marmosets (*Callithrix jacchus*) decreased their rates of self-scratching during a predator simulation (and after alarm calling began). While this appears to contradict our addition of self-scratching as part of stage two in a three-stage response to predation, Troisi et al. (1991) suggest that self-scratching increases only during moderate—as opposed to low or high—levels of anxiety. Levels of tolerance in various contexts likely determine when an individual experiences low, medium, or high levels of anxiety. Although there is little information about subtle behavioral responses at this second stage, they are likely to be marked by motivational conflict, i.e., uncertainty about whether to stay put to avoid detection, flee, or confront the predator. Given that displacement behaviors, including fear-related aggression toward conspecifics and SDBs, tend to be displayed during motivational conflict (Blurton Jones 1968; Van Lawick-Goodall: cited in Hinde 1974; Maestripieri et al. 1992), the increases in both types of response are likely to be seen when macaques are confronted by uncertainty. For example, tourists directly present within a group are likely perceived as riskier than when tourists are outside the group and easier to avoid. Maréchal et al. (2016) also showed this pattern of displacement behavior in habituated Barbary macaques, which appeared to depend on a trade-off between perceived risks vs. potential benefits (provisioned food) from tourists. This example has an important parallel to *M. nigra* in TNR for whom access to food from tourists, guides, garbage, and nearby crops could incentivize them to stay in this area despite heightened stress. It may also be that such behaviors help to mitigate/cope with the physiological effects of fear and stress (Higham et al. 2009). The final (3rd) stage of predation avoidance/defense usually includes either increased aggression toward or fleeing from predators (Beauchamp 2015).

In the present study, we found that when tourists were evident in the forest, but not within study groups, macaques in all three social groups showed evidence of inhibition of a wide range of behavior: affiliative, aggressive, sexual, and self-directed, responses that collectively could be considered partial or mild freezing responses. As such, it is possible that as unfamiliar tourists are heard in the forest, macaques practice vigilance to monitor the whereabouts of tourists and avoid detection. When tourists appear within groups, we found evidence of both inhibitions, for example, on days when large numbers of tourists appeared, and of motivational uncertainty in relation to risk; SDBs and conspecific aggression were both increased, consistent with a second stage of response to predators. Finally, although we did not record any instances of tourist-directed aggression or fleeing (stage 3 behavior), PB1 occasionally still fled from tourists approaching their group and, on rare occasions, macaques (in R2) attacked humans within groups without clear provocation (personal observation, 2015). Further exploration of these rare instances would be valuable.

Comparing our findings with those of earlier researchers of this population, it appears that as tourism in the park has grown over the years, *M. nigra* behavioral responses have changed. In the early days of observation and tourism in this population, macaques in the study groups typically fled from observers (MacKinnon and MacKinnon 1980) and later from groups of tourists larger than seven (Kinnaird and O'Brien 1996). During the current study, they only rarely fled when confronted with tourists or directed aggression toward tourists within groups. Overall, the results of this study suggest that primate groups exposed to tourism, even for decades, may not fully habituate to tourists. Although the groups now generally appear to tolerate the presence of large groups of tourists in the forest and within groups, our results challenge a common assumption among primate researchers and conservationists that, when long-term exposure to presumed benign anthropogenic influences such as tourism leads to apparent tolerance, habituation is complete. Rather, it appears as though tourists may still be perceived as sources of risk by such populations, inducing mild responses similar to predator avoidance. Whether these risk perceptions also lead to potentially harmful physiological stress responses, and their accompanying fitness effects, is not clear. However, several studies have shown evidence of increased glucocorticoids in response to tourists in other primates (Rangel-Negrín et al. 2014; Shutt et al. 2014; Cañadas Santiago et al. 2019).

Why males and females responded differently to familiar humans (researchers and tourist guides) is difficult to interpret. Males displayed less aggression when a greater number of researchers were present, while females displayed more aggression. Additionally, males displayed less affiliation when more guides were present, while females displayed less aggression. It may be that males and females differ in their risk perceptions of familiar humans based on their individual appearances or behavior rather than on (or in addition to) their numbers. There is also the possibility that, at least in the case of females, the presence of a greater number of guides reduces the potential threat of tourists. The differences between responses to researchers vs. guides may be twofold. First, researchers spend all day with macaque groups—from sunrise to sunset. This is a long period of time to have humans following and watching the group. While males may see them as a protective, familiar element, females (especially those with young infants) may not find their watchful presence as comforting. Secondly, MNP researchers undergo training and habituation to groups for several months before collecting data. Tourist behavior is not as controlled, and they lack the knowledge and understanding of primate behavior to make their presence less stressful. Notably, some MNP permanent research assistants also serve as guides. While not analyzed here, we recorded researcher IDs and guide names. With a deeper analysis, we may be able to uncover specific characteristics of familiar humans that play a role in the macaques' response, e.g. differences in gender, experience, and roles (researcher vs guide). Regardless, our results here urge caution for all primate field sites to review their protocols for number of researchers present at one time. If indeed researcher presence impacts primate behavior, this could be detrimental to group cohesion. Possible ways to mitigate these behavioral shifts could involve periodic assessments of monkey responses to researcher numbers and characteristics and could be paired with ongoing training.

While the results of this study appear to be reasonably consistent with a predator avoidance framework, they involved only a moderate number of subjects. Moreover, while the amount of time some subjects were observed in the presence of tourists was relatively short, the number of independent focal sessions themselves was high due to the short nature (~2 min) of each focal session. Future studies are needed with larger samples to validate them. Additionally, a more accurate measure of tourist numbers and attributes in the forest would be ideal. As of late 2016, TNR promoted their Ticket Master to full time and requested she keeps a daily log of tourist names, which guides attend which tourist groups, and total tourist counts. This includes tourists who only go to the beach, without intending to visit macaques or tarsiers specifically. Additionally, it would be useful to measure distances between focal macaques and tourist groups in order to examine responses to tourists at varying distances.

Future studies would also benefit by examining changes in behavior over time within a day. Is there a threshold number of tourists present at the same time that triggers a strong predator response, similar to the early Kinnaird and O'Brien study that uncovered a limit of seven? Additionally, it is possible that certain tourist characteristics (e.g., gender, age, national vs international) illicit stronger responses than others. Do the macaques respond more strongly to certain stimuli presented by some tourists, perhaps stimuli most closely associated with predation, or do they respond uniformly to all unfamiliar humans? While we know that many primates recognize different species of predators and respond adaptively with different behaviors (Cheney and Seyfarth 1981), evidence has shown that most primates have evolved more general predator avoidance tactics to specific stimuli, (e.g., unexpected sounds, moving shadows overhead, unexpected visual changes to the environment). Therefore, any organism that provides such stimuli is likely to elicit a predator avoidance/defensive response (see review Schel and Zuberbühler 2009).

Overall, it is important to note that SRBs themselves may be the result of a variety of causes. Untangling one direct cause is unlikely. However, it may be possible to demonstrate their relationship to stress physiology by complimenting these behavioral results with data on physiological responses to tourism, while also keeping in mind that glucocorticoids, including cortisol, are not only activated during periods of stress, but also play a primary function in energy mobilization and have numerous pleiotropic effects in vertebrates (MacDougall-Shackleton et al. 2019). Such a study should ideally relate behavioral and physiological responses to fitness-related measures, e.g., infant mortality rates, given that stress is, at its core, an adaptive response (Moberg 2000) that only becomes maladaptive under particular conditions (Sapolsky 1992). Similar to Beale and Monaghan (2004), all of the above could be combined into a comprehensive model of perceived predation risk and used as a framework for understanding the effects of tourist disturbance. Such a model should ultimately better enable conservation biologists and site managers to identify aspects of tourism and primate management in need of modification and thus bring tourism operations and human/animal conflict management practices in better alignment with their intended goals.

While a more comprehensive study is warranted, we tentatively recommend certain policies related to *M. nigra* viewing in Tangkoko. Both the numbers of guides and tourists should be limited. Guides should be encouraged to bring small groups incrementally into the forest. Additionally, it would be beneficial for paid guides or park rangers to monitor the beach area, as it is frequently visited by all three macaque groups. When tourists are in the forest, silence should be encouraged and feeding/touching discouraged. Tangkoko recently opened a "Visitor Center" at the entrance of the park. Paid staff could orient tourists on proper behavior around macaques, such as no eye contact, no rapid movements, and no touching the flora. Our recommendations leave room for additional job creation for the local village, providing new areas to educate visitors, protect the forest as a whole, and showcase this critically endangered species.

Acknowledgements We gratefully acknowledge all the guides, rangers, and people who work within and in connection with Tangkoko, specifically the Indonesian Government, Balai Konservasi Sumber Daya Alam (Conservation of Natural Resources in North Sulawesi), Kementerian Riset dan Teknologi (Indonesian Ministry of Research and Technology), Direktorat Jenderal Konservasi Sumber Daya Alam dan Ekosistem (Indonesian Directorate General of Nature Resources and Ecosystem Conservation), The American Indonesian Exchange Foundation, Macaca Nigra Project Team, Selamatkan Yaki, Tasikoki, and Tangkoko Conservation Education. D. Bertrand would personally like to thank her Anthropogenic Stress team, without whom, none of this data collection at Tangkoko would have been possible: Uni Suiah, Yuliana Sheza, Nur Aoliya, Mary Zuromskis, Kayla Wood, Rachel Sinsheimer. And she would also like to thank Dr. Lisa Danish and Dr. Jennifer Marciniak. A very special thank you goes to all of our funders: The National Science Foundation, Fulbright, Rufford Foundation, Chester Zoo, Nacey Maggioncalda Foundation, Primate Conservation Inc., University at Buffalo Anthropology Department, Nila T. Gnamm Research Fund, and the Mark Diamond Research Fund.

Literature Cited

Aguilar-Melo AR, Andresen E, Cristóbal-Azkarate J, Arroyo-Rodríguez V, Chavira R, Schondube J, Cuarón AD (2013) Behavioral and physiological responses to subgroup size and number of people in howler monkeys inhabiting a forest fragment used for nature-based tourism. Am J Primatol 75(11):1108–1116

Balasubramaniam KN, Marty PR, Arlet ME, Beisner BA, Kaburu SSK, Bliss-Moreau E, Kodandaramaiah U, McCowan B (2020a) Impact of anthropogenic factors on affiliative behaviors among bonnet macaques. Am J Phys Anthropol 171:704–717

Balasubramaniam KN, Marty PR, Samartino S, Sobrino A, Gill T, Ismail M, Saha R, Beisner BA, Beisner BA, Kaburu SSK, Bliss-Moreau E, Arlet ME, Ruppert N, Ismail A, Sah SAM, Mohan L, Rattan SK, Kodandaramaiah U, McCowan B (2020b) Impact of individual demographic and social factors on human–wildlife interactions: a comparative study of three macaque species. Sci Rep 10:21991. https://doi.org/10.1038/s41598-020-78881-3

Barros M, de Souza Silva MA, Huston JP, Tomaz C (2004) Multibehavioral analysis of fear and anxiety before, during, and after experimentally induced predatory stress in Callithrix penicillata. Pharmacol Biochem Behav 78(2):357–367

Beale CM, Monaghan P (2004) Human disturbance: people as predation-free predators? J Appl Ecol 41:335–343

Beauchamp G (2015). Animal vigilance: monitoring predators and competitors

Beehner JC, Bergman TJ (2017) The next step for stress research in primates: to identify relationships between glucocorticoid secretion and fitness. Horm Behav 91:68–83

Behie AM, Pavelka MS, Chapman CA (2010) Sources of variation in fecal cortisol levels in howler monkeys in Belize. Am J Primatol 72(7):600–606

Bejder L, Samuels A, Whitehead H, Finn H, Allen S (2009) Impact assessment research: use and misuse of habituation, sensitisation and tolerance in describing wildlife responses to anthropogenic stimuli. Mar Ecol Prog Ser 395:177–185

Bergman TJ, Beehner JC, Cheney DL, Seyfarth RM, Whitten PL (2005) Correlates of stress in free-ranging male chacma baboons, *Papio hamadryas ursinus*. Anim Behav 70(3):703–713

Blumstein D (2016) Habituation and sensitization: new thoughts about old ideas. Anim Behav 120:255–262

Blurton Jones NG (1968) Observations and experiments on causation of threat displays of the great tit (Parus major). Anim Behav Monogr 1(2):73–158

Cañadas Santiago S, Dias PAD, Garau S, Coyohua Fuentes A, Chavira Ramírez DR, Canales Espinosa D, Rangel Negrín A (2019) Behavioral and physiological stress responses to local spatial disturbance and human activities by howler monkeys at Los Tuxtlas, Mexico. Anim Conserv 23:297–306

Cavigelli SA, Caruso MJ (2015) Sex, social status and physiological stress in primates: the importance of social and glucocorticoid dynamics. Philos Trans R Soc B Biol Sci 370(1669):20140103

Chamove AS, Hosey GR, Schaetzel P (1988) Visitors excite primates in zoos. Zoo Biol 7:359–369

Cheney DL, Seyfarth RM (1981) Selective forces affecting the predator alarm calls of vervet monkeys. Behaviour 76(1–2):25–61

Clarke MR, Harrison RM, Didier ES (1996) Behavioral, immunological, and hormonal responses associated with social change in rhesus monkeys (*Macaca mulatta*). Am J Primatol 39:223–233

Coleman K (2012) Individual differences in temperament and behavioral management practices for nonhuman primates. Appl Anim Behav Sci 137:106–113

Danish L, Kerhoas D, Bertrand D, Febriyanti DY, Engelhardt A (2017) *Macaca nigra* (Desmarest, 1822). In: Schwitzer C, Mittermeier RA, Rylands AB, Chiozza F, Williamson EA, Macfie EJ, Wallis J, Cotton A (eds) Primates in peril: the world's 25 most endangered primates 2016–2018. IUCN SSC Primate Specialist Group (PSG)/International Primatological Society (IPS)/Conservation International (CI), and Bristol Zoological Society, Arlington, pp 64–67

Duboscq J, Micheletta J, Agil M, Hodges K, Thierry B, Engelhardt A (2013) Social tolerance in wild female crested macaques (*Macaca nigra*) in Tangkoko-Batuangus Nature Reserve, Sulawesi, Indonesia. Am J Primatol 75(4):361–375

Duboscq J, Romano V, Sueur C, MacIntosh AJJ (2016) Scratch that itch: revisiting links between self-directed behaviour and parasitological, social and environmental factors in a free-ranging primate. R Soc Open Sci 3:160571

Elston DA, Moss R, Boulinier T, Arrowsmith C, Lambin X (2001) Analysis of aggregation, a worked example: numbers of ticks on red grouse chicks. Parasitology 122(5):563–569

Engelhardt A, Perwitasari-Farajallah D (2008) Reproductive biology of Sulawesi crested black macaques (*Macaca nigra*). Folia Primatol 79:326

Engelhardt A, Heistermann M, Agil M, Perwitasari-Farajallah D, Higham J (2011) A despotic mating system in a socially tolerant primate, the crested macaque. Unpublished

Fichtel C, Kraus C, Ganswindt A, Heistermann M (2007) Influence of reproductive season and rank on fecal glucocorticoid levels in free-ranging male Verreaux's sifakas (*Propithecus verreauxi*). Horm Behav 51(5):640–648

Foley CAH, Papageorge S, Wasser SK (2001) Noninvasive stress and reproductive measures of social and ecological pressures in free-ranging African elephants. Conserv Biol 15:1134–1142

Frid A, Dill LM (2002) Human-caused disturbance stimuli as a form of predation risk. Conserv Ecol 6(1):11

Goodall J (1986) The chimpanzees of Gombe: patterns of behavior. Belknap Press of Harvard University Press, Cambridge, MA

Hahn-Holbrook J, Holt-Lunstad J, Holbrook C, Coyne SM, Lawson ET (2011) Maternal defense: breast feeding increases aggression by reducing stress. Psychol Sci 22(10):1288–1295

Hanson KT, Riley E (2017) Beyond neutrality: the human–primate interface during the habituation process. Int J Primatol 39:852–877

Higham JP, Maclarnon AM, Heistermann M, Ross C, Semple S (2009) Rates of self-directed behaviour and faecal glucocorticoid levels are not correlated in female wild olive baboons (*Papio hamadryas anubis*). Stress 12(6):526–532

Hilser H, Sampson H, Melfi V, Tasirin JS (2013) Sulawesi crested black macaque *Macaca nigra* species action plan: draft 1. Selamatkan Yaki/Pacific Institute, Manado

Hinde RA (1970) Animal behaviour: a synthesis of ethology and comparative psychology. McGraw-Hill, New York

Hinde RA (1974) Biological bases of human social behaviour. McGraw-Hill, New York

IUCN (2022) IUCN red list of threatened species. Version 2020–2. www.iucnredlist.org. Downloaded 18 Sept 2022

Jack KM, Lenz BB, Healan E, Rudman S, Schoof VA, Fedigan L (2008) The effects of observer presence on the behavior of *Cebus capucinus* in Costa Rica. Am J Primatol 70:490–494

Johns BG (1996) Responses of chimpanzees to habituation and tourism in the Kibale Forest, Uganda. Biol Conserv 78:257–262

Kinnaird MF, O'Brien TG (1996) Ecotourism in the Tangkoko DuaSudara Nature Reserve: opening Pandora's box? Oryx 30(1):65–73

Knight J (2009) Making wildlife viewable: habituation and attraction. Soc Anim 17:167–184

Kyes R, Iskandar E, Onibala J, Paputungan U, Laatung S, Huettmann F (2012) Long-term population survey of the Sulawesi Black Macaques (*Macaca nigra*) at Tangkoko Nature Reserve, North Sulawesi, Indonesia. Am J Primatol 75(1):88–94

Laméris DW, van Berlo E, Sterck EH, Bionda T, Kret ME (2020) Low relationship quality predicts scratch contagion during tense situations in orangutans (*Pongo pygmaeus*). Am J Primatol 82:e23138

Leasor H, Macgregor O (2014) Proboscis monkey tourism. In: Russon A, Wallis J (eds) Primate tourism: a tool for conservation? Cambridge University Press, Cambridge, pp 56–75

MacDougall-Shackleton SA, Bonier F, Romero LM, Moore IT (2019) Glucocorticoids and "stress" are not synonymous. Integr Org Biol 1(1):obz017

MacKinnon J, MacKinnon K (1980) The behavior of wild spectral tarsiers. Int J Primatol 1:361–379

Maestripieri D, Schino G, Aurel F, Troisi A (1992) A modest proposal: displacement activities as an indicator of emotions in primates. Anim Behav 44:967–979

Maréchal L, Semple S, Majolo B, Qarro M, Heistermann M, MacLarnon A (2011) Impacts of tourism on anxiety and physiological stress levels in wild male Barbary macaques. Biol Conserv 144(9):2188–2193

Maréchal L, Semple S, Majolo B, MacLarnon A (2016) Assessing the effects of tourist provisioning on the health of wild Barbary macaques in Morocco. PLoS One 11(5):e0155920

Martin JGA, Réale D (2008) Animal temperament and human disturbance: implications for the response of wildlife to tourism. Behav Process 77(1):66–72

Marty PR, Hodges K, Heistermann M, Agil M, Engelhardt A (2017a) Is social dispersal stressful? A study in male crested macaques (*Macaca nigra*). Horm Behav 87:62–68

Marty PR, Hodges K, Agil M, Engelhardt A (2017b) Alpha male replacements and delayed dispersal in crested macaques (*Macaca nigra*). Am J Primatol 79(7):e22448

Marty PR, Beisner B, Kaburu SSK, Balasubramaniam K, Bliss-Moreau E, Nadine Ruppert N, Sah SAM, Ismail A, Arlet ME, Atwill ER, McCowan B (2019) Time constraints imposed by anthropogenic environments alter social behaviour in longtailed macaques. Anim Behav 150:157–165

Matheson MD, Sheeran LK, Li JH, Wagner RS (2006) Tourist impact on Tibetan macaques. Anthrozoös 19(2):158–167

McDougall P (2012) Is passive observation of habituated animals truly passive?. J Ethol 30:219–223. https://doi.org/10.1007/s10164-011-0313-x

Mitchell G, Obradovich S, Herring F (1991) Threats to observers, keepers, visitors, and others by zoo mangabeys (*Cercocebus galeritus chrysogaster*). Primates 34(4):515–552

Moberg GP (2000) Biological response to stress: implications for animal welfare. In: Moberg GP, Mench JA (eds) The biology of human stress. CABI Publishing, Wallingford, pp 123–146

Muller K (1992) Underwater Indonesia. A guide to the world's greatest diving. Periplus Editions, Singapore

Muller MN, Wrangham RW (2004) Dominance, cortisol and stress in wild chimpanzees (Pan troglodytes schweinfurthii). Behav Ecol Sociobiol 55:332–340

Munck A, Guyre PM, Holbrook NI (1984) Physiological functions of glucocorticoids in stress and their relationship to pharmacological actions. Endocrinol Rev 5:25–44

Nakamichi M, Koyama N (2000) Intra-troop affiliative relationships of females with newborn infants in wild ring-tailed lemurs (*Lemur catta*). Am J Primatol 50:187–203

Natural Resources Management Project (1993) Ecotourism development in Bunaken National Park and North Sulawesi. Report no. 30 to Associates in Rural Development for Office of Agro-Enterprise and Environment, AID contract no. 497–0362

Neal SJ, Caine NG (2016) Scratching under positive and negative arousal in common marmosets (*Callithrix jacchus*). Am J Primatol 78:216–226

O'Brien TG, Kinnaird MF (1997) Behavior, diet, and movements of the Sulawesi Crested Black Macaque (*Macaca nigra*). Int J Primatol 18(3):321–351

Palacios JFG, Engelhardt A, Agil A, Hodges K, Bogia R, Waltert M (2011) Status of, and conservation recommendations for, the critically endangered crested black macaque Macaca nigra in Tangkoko, Indonesia. Oryx 46:290–297

Palagi E, Norscia I (2011) Scratching around stress: hierarchy and reconciliation make the difference in wild brown lemurs (*Eulemur fulvus*). Stress 14:93–97

Paulsen D (2009) The behavioral and physiological effects of ecotourism on the Sulawesi black macaques at the Tangkoko Nature Reserve, North Sulawesi, Indonesia. (Doctoral Dissertation), University of Washington

Pérez-Galicia S, Miranda-Anaya M, Canales-Espinosa D, Muñoz-Delgado J (2017) Visitor effect on the behavior of a group of spider monkeys (*Ateles geoffroyi*) maintained at an island in Lake Catemaco, Veracruz/Mexico. Zoo Biol 36:360–366

Pride RE (2005) Foraging success, agonism, and predator alarms: behavioral predictors of cortisol in *Lemur catta*. Int J Primatol 26:295–319

Rangel-Negrín A, Coyohua-Fuentes A, Chavira R, Canales-Espinosa D, Dias PAD (2014) Primates living outside protected habitats are more stressed: the case of black howler monkeys in the Yucatán Peninsula. PLoS One 9(11):e112329

Reed C, O'Brien TG, Kinnaird MF (1997) Male social behavior and dominance hierarchy in the Sulawesi Crested Black Macaque (*Macaca nigra*). Int J Primatol 18(2):247–260

Rhee S, Kitchener D, Brown T, Merrill R, Dilts R, Tighe S, USAID-Indonesia (2004) Report on Biodiversity and Tropical Forests in Indonesia. Submitted in Accordance with Foreign Assistance Act Sections 118/119, 1–316. Retrieved from http://www.irgltd.com/Resources/Publications/ANE/2004-02 Indonesia Biodiversity and Tropical Forest.pdf

Riley E (2010) The endemic seven: four decades of research on the Sulawesi macaque. Evol Anthropol 19:22–36

Roelofs K (2017) Freeze for action: neurobiological mechanisms in animal and human freezing. Philos Trans R Soc Lond Ser B Biol Sci 372(1718):20160206

Rosenbaum B, O'Brien TG, Kinnaird M, Supriatna J (1998) Population densities of Sulawesi crested black macaques (*Macaca nigra*) on Bacan and Sulawesi, Indonesia: effects of habitat disturbance and hunting. Am J Primatol 44:89–106

Rowell TE, Hinde RA (1963) Responses of rhesus monkeys to mildly stressful situations. Anim Behav 11:235–243

Sapolsky RM (1986) Endocrine and behavioral correlates of drought in wild olive baboons (*Papio anubis*). Am J Primatol 11:217–227

Sapolsky RM (1992) Stress, the aging brain, and the mechanisms of neuron death. MIT Press, London

Sapolsky RM (2005) The influence of social hierarchy on primate health. Science 308:648–652

Sapolsky R, Romero L, Munck A (2000) How do glucocorticoids influence stress responses? Integrating permissive, suppressive, stimulatory, and preparative actions. Endocr Rev 21(1):55–89

Sari IR (2013) Anthropogenic effects on habitat use, activity budget, and energy balance in Macaca nigra at Tangkoko Nature Reserve, North Sulawesi, Indonesia (Master's thesis). Georg-August University Göttingen

Schel A, Zuberbühler K (2009) Responses to leopards are independent of experience in Guereza colobus monkeys. Behaviour 146:1709–1737

Schino G, Perretta G, Taglioni AM, Monaco V, Troisi A (1996) Primate displacement activities as an ethnopharmacological model of anxiety. Anxiety 2:238–263

Sheriff MJ, Dantzer B, Delehanty B, Palme R, Boonstra R (2011) Measuring stress in wildlife: techniques for quantifying glucocorticoids. Oecologia 166(4):869–887

Shutt K, Heistermann M, Kasim A, Todd A, Kalousova B, Profosouva I, Petrzelkova K, Fuh T, Dicky J, Bopalanzognako J, Setchell JM (2014) Effects of habituation, research and ecotourism on faecal glucocorticoid metabolites in wild western lowland gorillas: implications for conservation management. Biol Conserv 172:72–79

Stanford CB (2002) Avoiding predators: expectations and evidence in primate antipredator behavior. Int J Primatol 23(4):741–757

Sugardjito J, Southwick CJ, Supriatna J, Kohlhass A, Baker S, Erwin J, Froehlich J, Lerche N (1989) Population survey of macaques in Northern Sulawesi. Am J Primatol 18:285–301

Supriatna J, Andayani N (2008) Macaca nigra. The IUCN red list of threatened species 2008: e.T12556A3357272

Supriatna J, Shekelle M, Fuad HA, Winarni NL, Dwiyahreni AA, Farid M, Sri Mariati, Margules C, Prakoso B, Zakaria Z (2020) Deforestation on the Indonesian island of Sulawesi and the loss of primate habitat. Glob Ecol Conserv 24:e01205

Troisi A (2002) Displacement activities as a behavioral measure of stress in nonhuman primates and human subjects. Stress 5:47–54

Troisi A, Schino G, D'Antoni M, Pandolfi N, Aureli F, D'Amato FR (1991) Scratching as a behavioral index of anxiety in macaque mothers. Behav Neural Biol 56(3):307–313

Tutin CEG, Fernandez M (1991) Responses of wild chimpanzees and gorillas to the arrival of primatologists: behaviour observed during habituation. In: Box HO (ed) Primate responses to environmental change. Springer, Dordrecht

Tyrrell M, Berman CM, Duboscq J, Agil M, Sutrisno T (2020) Avoidant social style among wild crested macaque males (Macaca nigra) in Tangkoko Nature Reserve, Sulawesi, Indonesia. Behaviour 157(5):451–491

Ulyan MJ, Burrows AE, Buzzell CA, Raghanti MA, Marcinkiewicz JL, Phillips KA (2006) The effects of predictable and unpredictable feeding schedules on the behavior and physiology of captive brown capuchins (Cebus apella). Appl Anim Behav Sci 101(1–2):154–160

Ventura R, Majolo B, Schino G, Hardie S (2005) Differential effects of ambient temperature and humidity on allogrooming, self-grooming, and scratching in wild Japanese macaques. Am J Phys Anthropol 126:453–457

Weingrill T, Gray DA, Barrett L, Henzi SP (2003) Fecal cortisol levels in free-ranging female chacmas baboons: relationship to dominance, reproductive state and environmental factors. Horm Behav 45:259–269

Whitehouse J, Micheletta J, Waller B (2017) Stress behaviours buffer macaques from aggression. Sci Rep 7:11083

Williamson E, Feistner A (2011) Habituating primates: processes, techniques, variables and ethics. In: Setchell J, Curtis D (eds) Field and laboratory methods in primatology: a practical guide. Cambridge University Press, Cambridge, pp 33–50

Wittig RM, Crockford C, Lehmann J, Whitten PL, Seyfarth RM, Cheney DL (2008) Focused grooming networks and stress alleviation in wild female baboons. Horm Behav 54(1):170–177

Woods JM, Ross SR, Cronin KA (2019) The social rank of zoo-housed Japanese Macaques is a predictor of visitor-directed aggression. Animals 9:316

Yamagiwa J (2011) Ecological anthropology and primatology: fieldwork practices and mutual benefits. In: MacClancey J, Fuentes A (eds) Centralizing fieldwork: critical perspectives from primatology, biological and social anthropology. Berghahn Books, pp 84–103

Chapter 5
The Effect of Tourism on a Nocturnal Primate, *Tarsius Spectrum*, in Indonesia

Sharon L. Gursky

Abstract The primary goal of conservation is to maintain biological diversity. Since the nineteenth century, public areas such as national parks and nature reserves provide the main mechanism by which conservationists strive to maintain biological diversity. The biggest impediment faced by conservationists to successfully preserve biodiversity is insufficient financial resources. In response to the shortage of funds, conservationists have developed the concept of "sustainable ecotourism" to fund conservation activities. Sustainable ecotourism involves people paying to visit fragile, pristine, and relatively undisturbed natural areas. The goal of this study was to explore the effect of tourism on Gursky's Spectral Tarsier (*Tarsius spectrumgurskyae*). The results of this preliminary study clearly show that the presence and behavior of tourists clearly affect the behavior of the tarsiers. Tarsier groups that were exposed to tourists departed their sleeping tree at significantly higher heights, emitted significantly more audible alarm calls prior to departing their sleeping site, emitted significantly more ultrasonic alarm calls prior to departing their sleeping site, left their sleeping site significantly later, and were more likely to not return to their main sleeping site than did groups that were not exposed to the tourists. These results show that even when wildlife viewing is carried out exclusively by qualified and trained guides, tourism led to substantial changes in behavior of the viewed tarsiers.

Keywords Prosimian · Strepsirrhine · Sulawesi · ALAN · Indonesia · Tarsier

Author Contribution: This project was designed and conducted solely by the author.

S. L. Gursky (✉)
Department of Anthropology, MS 4352, Texas A&M University, College Station, TX, USA
e-mail: gursky@tamu.edu

5.1   Introduction

The primary goal of conservation is to maintain biological diversity (Buckley et al. 2016; Pereira et al. 2010; Butchart and Bird 2010; Barnosky et al. 2010; Pimm et al. 2014). Biological diversity (aka biodiversity) refers to the biological variation of life on earth. Biodiversity is not evenly distributed throughout the world but is greatest between the Tropics of Cancer and Tropics of Capricorn, despite the fact that tropical ecosystems represent less than 10% of the earth's surface. The uneven distribution of biodiversity is also expressed in that within the tropics there are currently 36 recognized biodiversity hotspots (IUCN 2020); areas that support nearly half of the world's endemic mammal species while comprising less than 2.5% of the land on earth. Human modification of earth's ecosystems has and continues to alter the earth's biodiversity. Recognizing that biodiversity not only influences environmental conditions, but also provides our fuel, biomaterials, biofuels, pollination, genetic resources, and many other benefits (Hillebrand et al. 2017), for decades conservationists have striven to protect this resource.

Since the nineteenth century, public areas such as national parks and nature reserves provide the main mechanism by which conservationists strive to maintain biological diversity (Watson et al. 2014; Rands et al. 2010). According to the IUCN (2016), protected areas not only conserve biodiversity, but also contribute to national and local economies and are the foundation for sustainable livelihoods in many communities. Although globally the total number of protected areas have tripled over the last several decades, they currently only account for less than 15% of the world's terrestrial area (IUCN 2016). This is less than the 17% goal by 2020 that was set by the Convention on Biological Diversity (Juffe-Bignoli 2014). This number is quite variable by country with India having a mere 5.9% of its land in protected areas, Indonesia 12.7% of its land, and Cambodia 26% of their land in protected areas (Index Mundi 2020).

Despite the best efforts of conservationists worldwide, species extinctions continue to occur at an alarming rate worldwide (Barnosky et al. 2011; Pimm et al. 2014). As an example, note that in 2019 alone we lost three bird species (**Alagoas foliage-gleaner,** *Philydor novaesi;* **Cryptic treehunter,** *Cichlocolaptes mazarbarnetti;* Poo-uli, *Melamprosops phaeosoma*), a frog (**Corquin robber frog,** *Craugastor anciano),* a shark **Lost shark** (*Carcharhinus obsolerus*), a snail (Hawaiian tree snail, *Achatinella apexfulva*), a skink (**Boulenger's speckled skink,** *Oligosoma infrapunctatum*), a rat (Bramble Cay mosaic tailed rat, *Melomys rubicola*), and numerous fish including the **Chinese paddlefish** (*Psephurus gladius*), **Cunning silverside** (*Atherinella callida*), and the **Catarina pupfish** (*Megupsilon aporus*). All of these species were declared extinct in one year (IUCN 2020). These changes in biodiversity have led to what is often called a biodiversity crisis, with warnings that current rates of extinctions are exceptionally high (Mace et al. 2005; Pimm et al. 2014), and suggestive of an impending global mass extinction phenomenon (Barnosky et al. 2012).

The biggest impediment faced by conservationists to successfully preserve biodiversity is insufficient financial resources (Butchart et al. 2012). According to one analysis, protecting all the world's threatened species will cost around four billion a year (McCarthy et al. 2012). In response to the shortage of funds, conservationists have developed the concept of "sustainable ecotourism" to fund conservation activities (McAfee 1999). Sustainable ecotourism involves people paying to visit fragile, pristine, and relatively undisturbed natural areas. These trips are intended to be low-impact and are supposed to be small scale in comparison to standard commercial mass tourism. The main purposes of sustainable ecotourism is to educate the traveler, to provide funds for ecological conservation, to directly benefit the economic development and political empowerment of local communities, as well as to foster respect for different cultures and for human rights (Higginbottom 2004). Since the 1980s, ecotourism has been considered a critical endeavor by environmentalists so that future generations may experience destinations relatively untouched by human intervention (Higginbottom 2004).

Currently, conservation efforts worldwide rely increasingly on ecotourism for financial and political support. National parks agencies worldwide receive as much as 84% of their funding from ecotourism (Buckley et al. 2016). Ecotourism has become one of the fastest-growing sectors of the tourism industry, growing annually by 10–15% worldwide (UNWTO 2017). Just as one example, in 2008, 13 million people took part in whale watching, generating an expenditure of $2.1 billion (O'Connor et al. 2009).

While ecotourism funds the majority of conservation activities, there have been very few studies that explore the effects of ecotourism on the habitat and species that they are designed to protect. The few studies that have been conducted suggest that "sustainable ecotourism" may be contributing to the extinction crisis (Lusseau and Bejder 2007; Bejder et al. 2006; Christiansen et al. 2015; McClung et al. 2004; Pirotta et al. 2014; Watson et al. 2014). Ecotourism puts extra pressures on the local environment and necessitates the development of additional infrastructure and amenities (Liu et al. 2012). The construction of water treatment plants, sanitation facilities, and lodges comes with the exploitation of nonrenewable energy sources and the utilization of already limited local resources (Vivanco 2002). The conversion of natural land to such tourist infrastructure is implicated in deforestation and habitat deterioration of butterflies in Mexico and squirrel monkeys in Costa Rica (Isaacs 2000). Aside from environmental degradation with tourist infrastructure, population pressures from ecotourism also leave behind garbage and pollution associated with the Western lifestyle (McLaren 1998).

As ecotourism is used and advocated more widely in conservation, quantifying its effects on the animals it is designed to protect has become correspondingly urgent. The goal of this study was to explore the effect of tourism on Gursky's Spectral Tarsier (*Tarsius spectrumgurskyae*). Gursky's Spectral Tarsier is endemic to Tangkoko Nature Reserve. Due to the habituation of these small nocturnal primates by the PI, the development of roads leading to the nature reserve, discussion of visibility in travel guides such as Lonely Planet, ecotourism to see these animals has been booming. Whereas in the late 1990s there were fewer than 100 tourists per year, there are now thousands of tourists (domestic and international) per year visiting the field site. Just as noticeable as the large increase in the tourists is how the tourists view the tarsiers. They come in very large groups (>10), armed with large, bright flashlights which they aim directly in the eyes of the animals. They get extremely close to the sleeping site and to the animals to maximize their photo opportunities. With such a massive influx in people using very invasive techniques to view the tarsiers each night necessitates asking whether the tourism is affecting the animal's behavior and general well-being? The goal of this preliminary research project was to explore how tourism affects specific aspects of the well-being of the tarsiers.

5.2 Methods

The island of Sulawesi is the largest island of the biogeographical region of Wallacea, a transition zone between the Australian and Asian zoogeographical regions (Audley-Charles 1981; Whitmore 1987). Consequently, Sulawesi shows a blend of Asian and Australian elements in its fauna and flora. Sulawesi also exhibits very high levels of endemic species. Sulawesi is the home of more than 260 bird species, 80 of which are endemic. Of the 127 indigenous mammals, 79 (62%) are endemic (Musser 1987). Endemic species include: anoa *Bubalus depressicornis,* macaque *Macaca nigra,* spectral tarsier *Tarsius spectrum,* and babirusa *Babyrousa babirousa.* I conducted this study at Tangkoko Nature Reserve on the easternmost tip of the northern arm of Sulawesi. The reserve exhibits a full range of forest types, including beach formation forest, lowland forests, submontane forests, and mossy cloud forests on the summits of the Tangkoko crater (MacKinnon and MacKinnon 1980; Whitten et al. 1987; Gursky 1997). The reserve is far from pristine due to heavy selective logging and encroaching gardens along its borders. The forest canopy is very discontinuous and contains a high proportion of *Ficus* trees (Gursky 1997, 1998). Rainfall averaged approximately 2,300 m annually, with most rainfall occurring between November and April (World Wildlife Fund 1980; Gursky 1997). Additional details concerning the habitat type at Tangkoko Nature Reserve can be found in Gursky (1997, 2007).

The following procedures were used to locate groups that had not previously been observed by the tourists. Prior to dawn, my field assistant and I would stand on the periphery of a 1-ha plot within the trail system at Tangkoko Nature Reserve. As the tarsiers returned to their sleeping site, or at their sleeping site, they gave loud vocal calls for 3-5 min that could be heard from 300 to 500 m (MacKinnon and MacKinnon 1980; Niemitz 1984). Groups that were heard vocalizing were then followed to their sleeping site. My field assistant and I then returned to the sleeping site prior to dawn the following morning to count the number of individuals leaving each sleeping tree as well as their relative age and sex.Four tarsier groups were located from the early morning audible vocalizations, each individual tarsier emits upon returning to its sleeping site. These vocalizations were given for 3–5 min and were heard from 300 to 400 meters. Five nights each week, my Indonesian field assistants and I observed the behavior of four tarsier groups as they were entering and leaving their relative sleeping sites. These times were chosen as this is when the park guides bring tourists to view the tarsiers. With the assistance of the park guards and guides, two groups were exposed to tourists regularly and two groups received no tourist visits. To minimize the effect of the research observers on the behavior of the tarsiers that were not exposed to tourists, blinds were created near the sleeping sites behind which my assistants and I remained. From behind the blind, my field assistants and I recorded the time that tarsiers left their sleeping site, the number of alarm calls (audible and ultrasonic) the tarsiers emitted prior to leaving their sleep site, whether or not the tarsiers returned to their sleeping tree, the height at which they departed from their sleeping tree as well as the number of tourists, the number of flashlights, and the mean distance of the tourists to the tarsier's sleeping tree.

5.3 Results

Over the four months of this study (September–December), which occurred after the height of the tourist season (June–August), a total of 685 tourists were observed over 87 nights at the two tarsier sleeping trees that were exposed to tourists. The range of tourists at the two sleeping trees ranged from 1 to 12 per night. This amounts to a mean of 3.9 (SD 2.8) tourists, per tarsier group, per night.

The mean height that the four tarsier groups left their sleeping site is 7.6 m (S.D. 2.1). However, the mean height that the tarsiers left their sleeping tree was affected by whether they were experiencing tourism (Fig. 5.1). Groups that did not experience tourism departed their sleeping tree at significantly lower heights (mean 4.62 m; S.D. 0.85), whereas groups that experienced tourism departed their sleeping tree at significantly higher heights (mean 10.35 m; S.D. 1.23) ($t = 12.301$; $p = 0.001$).

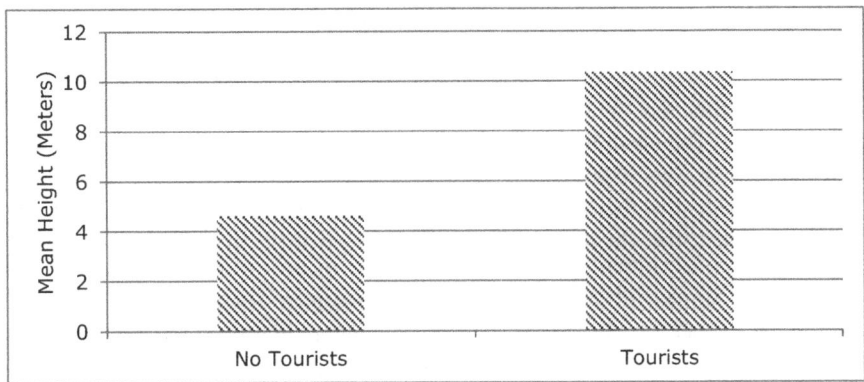

Fig. 5.1 The mean height that the spectral tarsiers left their sleeping sites when no tourists visited their sleeping tree compared to nights when tourists visited their sleeping site

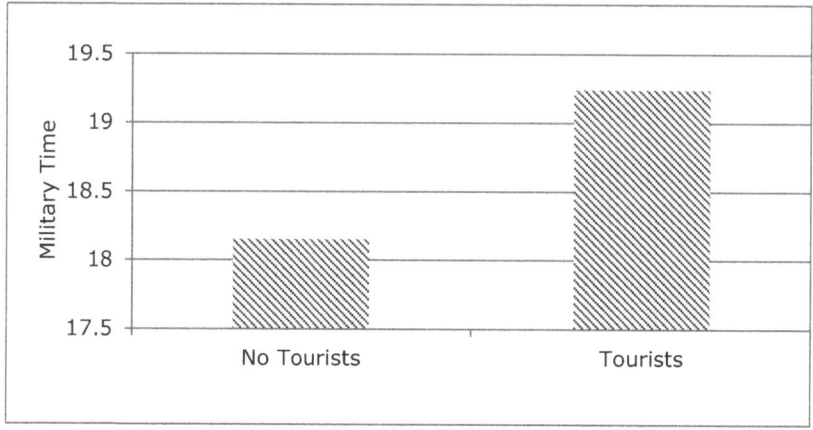

Fig. 5.2 The mean time that the tarsiers exited their sleeping trees when there were tourists and nights when there were no tourists

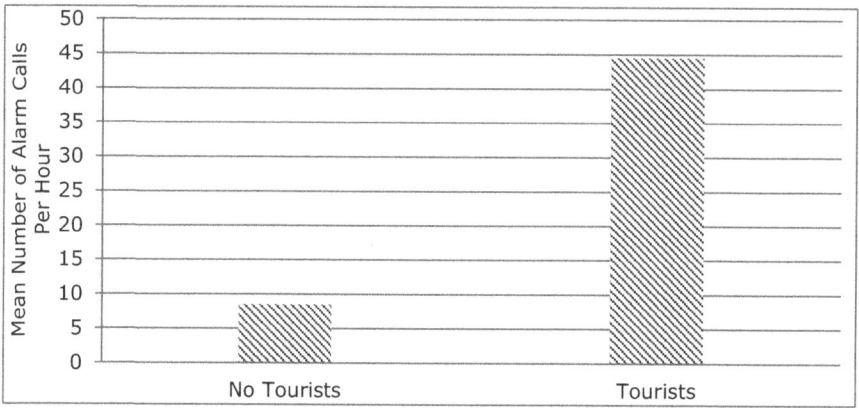

Fig. 5.3 The mean number of alarm calls emitted per hour when no tourists visited their sleeping tree compared to nights when tourists visited their sleeping site

The mean time that the four tarsier groups left their sleeping site averaged 18:53 (S.D. 48 min). The mean time that the tarsiers left their sleeping tree was affected by whether they were experiencing tourism (Fig. 5.2). Groups that did not experience tourism departed their sleeping tree significantly earlier at 18:15 (S.D. 21 min), whereas groups that experienced tourism departed their sleeping tree at 19:24 (S.D. 33 min) (t = 24.302; p = 0.001).

The mean number of alarm calls that the four tarsier groups emitted averaged 26.3 per hour (S.D. 14.85). However, the mean number of alarm calls that the tarsiers emitted were affected by whether they were experiencing tourism (Fig. 5.3). Groups that did not experience tourism emitted significantly fewer alarm calls per hour prior to departing their sleeping tree (mean = 8.5; S.D. 1.45), whereas groups

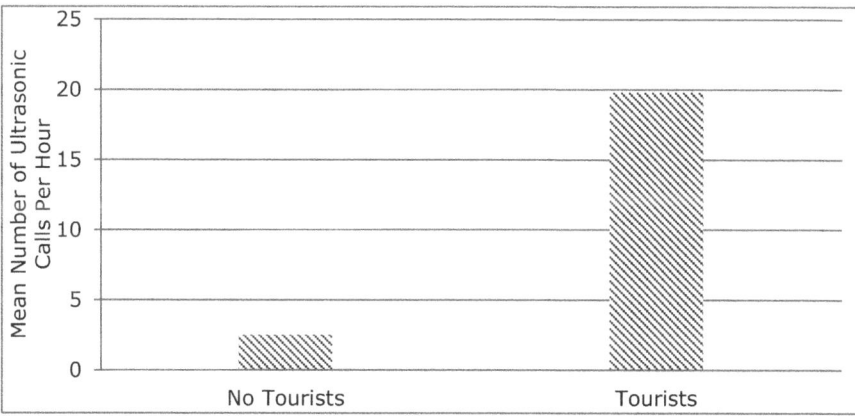

Fig. 5.4 Mean number of ultrasonic vocalizations emitted per hour on nights when tourists were present and nights when there were no tourists

that experienced tourism emitted significantly more alarm calls per hour prior to departing their sleeping tree (mean 44.5; S.D. = 18.23) (t = 7.481; p = 0.001).

In addition to audible alarm calls, mean number of ultrasonic calls that the four tarsier groups emitted averaged 13.1 per hour (S.D. 4.29). However, the mean number of ultrasonic calls that the tarsiers emitted were affected by whether they were experiencing tourism (Fig. 5.4). Groups that did not experience tourism emitted significantly fewer ultrasonic calls per hour prior to departing their sleeping tree (mean = 2.5; S.D. 0.76), whereas groups that experienced tourism emitted significantly more ultrasonic calls per hour prior to departing their sleeping tree (mean 19.8; S.D. = 10.63) (t = 6.150; p = 0.001). Throughout this preliminary study, the tarsiers returned to their main sleeping tree on 92% of the mornings (n = 80). The tarsier groups that did not experience tourism always returned to their main sleeping tree while the tarsier groups that were experiencing tourism occasionally did not return to their main sleeping site 16% (n = 7).

In addition to observing how the tarsiers responded to the tourists, ad libitum observations of the tourists were also recorded. These observations demonstrated evidence that individuals were observed feeding the tarsiers crickets (n = 3) in attempts to get better photographs as well as banging on the tree (n = 2) when no tarsiers were observed leaving the sleeping site. Tourists were also observed shining their flashlights up the tree (n = 26) and attempting to hand capture a tarsier (n = 1).

5.4   Conclusions

The results of this preliminary study clearly show that the presence and behavior of tourists clearly affect the behavior of the tarsiers. Tarsier groups that were exposed to tourists departed their sleeping tree at significantly higher heights, emitted significantly more audible alarm calls prior to departing their sleeping site, emitted significantly more ultrasonic alarm calls prior to departing their sleeping site, left their sleeping site significantly later, and were more likely to not return to their main sleeping site than did groups that were not exposed to the tourists. These results show that even when wildlife viewing is carried out exclusively by qualified and trained guides, tourism led to substantial changes in behavior of the viewed tarsiers.

While the preliminary nature of this research makes it impossible to quantitatively ascertain the ultimate fitness effects of these behavioral changes, the impacts on population viability can be inferred qualitatively. To begin with, it is known that increased energetic challenges, either as added traveling costs or reduced foraging opportunities, can lead to reduced reproductive success for individuals (Crofoot 2013; Boinski et al. 2002; Borries et al. 2008). For example, Dunn et al. (2013) observed that howlers that spend more time traveling to locate food had lower reproductive success and lower survival rates. If such challenges affect the tarsiers too frequently, then individuals will shift into long-term avoidance strategies when possible by avoiding the degraded areas. The fact that the tarsiers that are being observed by tourists were significantly more likely to change their sleeping site supports/ suggests the fact that the impact of the tourists is sometimes great enough to decrease the importance of their sleeping tree. This avoidance behavior has been observed in other organisms. For example, bottlenose dolphins (*Tursiops truncatus*) have been observed avoiding important foraging areas when motorboat traffic was high (Lusseau 2006). Another primate, the pygmy marmoset (*Cebuella pygmaea*) is also known to move to the upper canopy in areas disturbed by ecotourists. However, in less disturbed areas (i.e., fewer ecotourists), the pygmy marmoset uses their preferred lower forest levels (de la Torre et al. 2000). Similarly, the reduced time available each night to forage resulting from the delayed departure from sleeping tree because of tourists shining lights in their eyes, may also cause the tarsiers to devalue the territory. The decision to stay or leave a habitat that has been degraded by tourists will have to be evaluated individually and decided in terms of their individual costs and benefits. Individuals that cannot leave degraded habitat will have reduced fitness potentially leading to reduced reproductive success. While there have not been a huge number of studies showing behavioral changes due to tourism, the body of literature is certainly growing. A wide variety of short-term effects have been detected on many cetacean species being observed by boats including changes in their respiration patterns as well as changes in their path direction as an avoidance strategy (Lusseau 2006; Frid and Dill 2002).

Given the substantial behavioral changes that have been observed in tarsiers due to ecotourism, it is obvious that changes need to be implemented. Ending ecotourism for the tarsiers would probably negatively affect both the tarsiers and the people. The local guides are financially dependent on the salaries they obtain from guiding and removing their salary would just make them more dependent on forest resources. It is also well known that the presence of tourists and researchers hinders exploitation of the forest resources, thereby protecting the tarsiers and the forest resources they are dependent upon. Instead, a more moderate approach needs to be taken whereby retraining of guides need to be conducted annually. At Tangkoko

Nature Reserve, there used to be a formal "Code of Conduct" for guides, but it is no longer utilized. This is because tourist satisfaction is usually the driving goal of the guides. As tourist satisfaction increases so do the tips by the tourists. It is therefore recommended that a Code of Conduct be re-implemented at Tangkoko that incorporates a minimum distance to be kept from the tarsiers, a maximum number of tourists to visit each tarsier group, the use of infrared filters on flashlights for viewing the animals, and training of the guides to remember that their driving goal is not to maximize tourist satisfaction and thus maximize their tips, but to protect the tarsiers and their environment; increasing the cost of the tickets to include mandatory payments to the guides and prohibiting tipping by the tourists. Hopefully, the implementation of these strategies will minimize the effect of tourism on the tarsier population.

Acknowledgments I thank the Indonesian Institute of Sciences, the Directorate General for Nature Preservation and Forest Protection in Manado, Bitung, Tangkok, and Jakarta, SOSPOL, POLRI, the University of Indonesia, Jatna Supriatna for his sponsorship while in Indonesia. Special thanks for my field assistants Ben and Felix for their help in collecting the data.

Conflict of Interest Statement   The author has no conflict of interest to declare.

Funding Source   This work was funded by the L.S.B. Leakey Foundation, Texas A&M University, Conservation International Global Wildlife Fund and Primate Conservation Inc.

Statement of Ethics   This research complied with the ethical standards in the treatment of animals required by IACUC at Texas A&M University and with the laws of Indonesia.

References

Audley-Charles M (1981) Geological history of the region of Wallace's line. In: Whitmore TC (ed) Wallace's line and plate tectonics. Clarendon Press, Oxford

Barnosky A, Walpole M, Collen B, van Strien A, Scharlemann J, Watson R (2010) Global biodiversity: indicators of recent declines. Science 328:1164–1168

Barnosky A, Hadly E, Bascompte J, Berlow E, Brown J, Fortelius M et al (2012) Approaching a state shift in Earth's biosphere. Nature 486:52–58

Barnosky A, Matzke N, Tomiya S, Wogan G, Swartz B, Quental T et al (2011) Has the Earth's sixth mass extinction already arrived? Nature 471:51–57

Bejder L, Samuels A, Whitehead H, Gales N, Mann J, Connor R, Heithaus M, Watson-Capps J, Flaherty C, Krützen M (2006) Decline in relative abundance of bottlenose dolphins exposed to long-term disturbance. Conserv Biol 20:1791–1798

Boinski S, Sughrue M, Selvaggi L, Cropp S (2002) An expanded test of the ecological model of primate social evolution: competitive regimes and female bonding in three species of squirrel monkeys (Saimiri oerstedii, S. boliviensis, and S. sciureus). Behaviour 139:227–261

Borries C, Larney E, Ku A, Ossi K, Koenig A (2008) Costs of group size: lower developmental and reproductive rates in larger groups of leaf monkeys. Behav Ecol 19:1186–1191

Buckley R, Morrison C, Castley J (2016) Net effects of ecotourism on threatened species survival. PLoS One 11(2):e0147988

Butchart S, Bird J (2010) Data-deficient birds on the Iucn Red List: What don't we know and why does it matter? Biol Conserv 143:239–247

Butchart S, Scharlemann J, Evans M (2012) Protecting important sites for biodiversity contributes to meeting global conservation targets. PLoS One 7(3):e32529

Christiansen F, Bertulli D, Lusseau D (2015) Estimating cumulative exposure of wildlife to non-lethal disturbance using spatially explicit capture-recapture models. J Wildl Manag 79:311–324

Crofoot M (2013) The cost of defeat: capuchin groups travel further, faster and later after losing conflicts with neighbors. Am J Phys Anthropol 152:79–85

de la Torre S, Snowdon C, Bejarano M (2000) Effects of human activities on wild pygmy marmosets in Ecuarian Amazonia. Biol Conserv 94:153–163

Dunn J, Cristobal-Azkarate J, Schulte-Herbrueggen B (2013) Travel time predicts fecal glucocorticoid levels in free ranging howlers, Alouatta palliata. Int J Primatol 34:246–259

Frid A, Dill L (2002) Human caused disturbance stimuli as a form of predation risk. Conserv Ecol 6(1) https://www.jstor.org/stable/e26271845

Gursky S (1997) Modeling maternal time budgets: the impact of lactation and gestation on the behavior of the spectral tarsier, Tarsius spectrum. Ph.D. Dissertation, SUNY-Stony Brook

Gursky S (1998) Conservation status of the spectral tarsier *Tarsius spectrum*: population density and home range size. Folia Primatol 69:191–203

Gursky S (2007) The spectral tarsiers. In: Vasey N, Sussman R (eds) *Primate field studies* series. Prentice Hall, New York

Higginbottom K (2004) Wildlife tourism: an introduction. In: Higginbottom K (ed) Wildlife tourism: impacts, management and planning. Common Ground Publishing, Altona, Australia, pp 1–11

Hillebrand H, Blasius B, Borer E, Chase J, Downing J, Eriksson B, Ryabov A (2017) Biodiversity change is uncoupled from species richness trends – consequences for conservation and monitoring. J Appl Ecol 55:169–184

Isaacs 2000

IUCN (2016) Red data book. IUCN Press, New York

IUCN (2020) Red data book. IUCN Press, New York

Juffe-Bignoli D (2014) Protected planet report 2014. https://www.sogou.com/link?url=hedJjaC291NT6x-BmTJ9rWjL_M3bYAEkaQHXYjyuWmkmfBgxL9IPBvdtGKGggy6c

Liu W, Vogt CA, Luo J, He G, Frank KA, Liu J (2012) Drivers and socioeconomic impacts of tourism participation in protected areas. PLoS One 7(4):e35420

Lusseau D (2006) Unsustainable dolphin watching tourism in Fiorland. NewZealand Tour Mar Environ 3:173–178

Lusseau D, Bejder M (2007) Desperately seeking sustainability: a model to address the development and long-term management of tourist interactions with marine mammals. Report SC/59/WW4 submitted to the Scientific Committee of the International Whaling Commission. In: 59th annual meeting of the International Whaling Commission (IWC), Anchorage, Alaska, USA, pp 8–18

Mace G, Masundire H, Baillie J, Ricketts T, Brooks T, Hoffmann M, Williams P (2005) Biodiversity. In: Hassan H, Scholes R, Ash N (eds) Biodiversity in ecosystems and human wellbeing: current state and trends. Island Press, Washington, DC, pp 77–122

MacKinnon J, MacKinnon K (1980) The behavior of wild spectral tarsiers. Intl J Primatol 1:361–379

McAfee K (1999) Selling nature to save it? Biodiversity and green developmentalism. Environ Plan D Soc Space 17:133–154

McCarthy P, Jörn D, Scharlemann P, Buchanan G, Balmford A, Green J, Bennun L, Burgess N, Fishpool L, Garnett S, Leonard D, Maloney R, Morling P, Schaefer H, Symes H, Wiedenfeld A,

Butchart S (2012) Financial costs of meeting global biodiversity conservation targets: current spending and unmet needs. Science 3138:946–949

McClung M, Seddon P, Massaro M, Setiawan A (2004) Nature-based tourism impacts on yellow-eyed penguins *Megadyptes antipodes*: does unregulated visitor access affect fledging weight and juvenile survival? Biological Conservation 119:279–285

McLaren D (1998) Rethinking tourism and ecotravel: the paving of paradise and what you can do to stop it. Kumarian Press, Connecticut

Index Mundi (2020) Country Facts. Indexmundi.com

Musser G (1987) The mammals of Sulawesi. In: Whitmore TC (ed) Biogeographical evolution of the malay archipelago. Clarendon Press, Oxford

Niemitz C (1984) Biology of tarsiers. Fischer, Stuttgart

O'Connor S, Campbell R, Cortez H, Knowles T (2009) Whale watching worldwide: tourism numbers, expenditures and expanding economic benefits; IWC 61/14. International Fund for Animal Welfare, Yarmouth, MA, p 23

Pereira H, Leadley P, Proença V, Alkemade R, Scharlemann J, Fernandez-Manjarrés J, Araújo M, Balvanera P, Biggs R, Cheung W et al (2010) Scenarios for global biodiversity in the twenty-first century. Science 330:1496

Pimm S, Jenkins C, Abell R, Brooks T, Gittleman J, Joppa L, Sexton J (2014) The biodiversity of species and their rates of extinction, distribution, and protection. Science 344:987

Pirotta E, Thompson P, Cheney B, Donovan C, Lusseau D (2014) Estimating spatial, temporal and individual variability in dolphin cumulative exposure to boat traffic using spatially explicit capture-recapture methods. Anim Conserv 17:1–12

Rands M, Adams W, Bennun L, Butchart S, Clements A, Coomes D, Entwistle A, Hodge I, Kapos V, Scharlemann J, Sutherland W, Vira B (2010) Biodiversity conservation: challenges beyond. Science 329:1298–1303

UNWTO. UNWTO Tourism Highlights (2017) Available online: https://www.e-unwto.org/doi/pdf/10.18111/9789284419029. Accessed 13 Jan 2018

Vivanco L (2002) The international year of ecotourism in an age of uncertainty. The Ecologist

Watson J, Dudley N, Segan D, Hockings M (2014) The performance and potential of.protected areas. Nature 515(7525):67–73

Whitten T, Mustafa M, Henderson G (1987) The ecology of sulawesi. Gadjah Mada University Press, Yogyakarta

Whitmore TC (1987) Tropical rain forests of the far east. Oxford University Press, Oxford

World Wildlife Fund (1980) Cagar Alam Gunung Tangkoko Dua Saudara Sulawesi Utara Management Plan 1981–1986. World Wildlife Fund, Bogor Indonesia

Chapter 6
Javan Gibbon Tourism: A Review from West and Central Java Initiatives

Jatna Supriatna, Anton Ario, and Arif Setiawan

Abstract The goal of this paper is to review the tourism activities relating to the Javan gibbon. The tourism activities occur at three different sites in Java: Bodogol Education Center in the Gunung Gede National Park, Gunung Halimun, Salak National Park in West Java, and Swara Owa in Central Java. In Gunung Haliman, gibbon tourism is also community based and is integrated into cultural tourists' activities. In Gunung Gede, wildlife tourism focuses on education with multiple activities and programs for individuals of various educational levels. In Swara Owa, wildlife tourism centers on showing tourists the gibbons that are living under shade grown coffee, providing a beautiful example of how the needs of the local people and the necessity for conservation can be successfully implemented. Together the three gibbon wildlife tourism projects bring together, community, conservation, education, and development.

Keywords Javan gibbon · Ecotourism · Primates · Java · Indonesia

6.1   Introduction

Java represents the natural south-eastern limit of primate distribution in Asia. Some of the primates living in Sumatra and Kalimantan have become extinct on Java, for example, pig-tailed macaques, orangutans, and tarsiers. Java has four species of

J. Supriatna (✉)
Department of Biology, Faculty of Mathematics and Sciences, University of Indonesia, Depok, Indonesia

A. Ario
Conservation International Indonesia, Jakarta, Indonesia

A. Setiawan
SwaraOwa, Central Java, Desa Sokokembang, Petungkriyo, Pekalongan, Indonesia

© The Author(s), under exclusive license to Springer Nature Switzerland AG 2022   93
S. L. Gursky et al. (eds.), *Ecotourism and Indonesia's Primates*, Developments in Primatology: Progress and Prospects, https://doi.org/10.1007/978-3-031-14919-1_6

endemic primates, which are all classified as endangered: the Javan gibbon (*Hylobates moloch*), the grizzled leaf monkey (Presbytis comata), the West Javan langur (*Trachypithecus mauritius*), and the Javan slow loris (*Nycticebus javanicus*) (Roos et al. 2014). There are two other nonendemic primates: the Javan langur (*Trachypithecus auratus*), which is found in East Java to Bali and Lombok islands and the long-tailed macaque (*Macaca fascicularis*), which is distributed widely in Sundaland and introduced to several islands in the Nusa Tenggara archipelago (Roos et al. 2014; Supriatna 2019a, b).

Primates use forest as habitat, and most of the remaining forests on Java are in rugged mountainous areas or in national parks and other protected areas. There is now less than 6% of the original forest left on the island (Supriatna et al. 1994; Miettinen et al. 2011). Forest habitat has been converted into settlements and agricultural land. In addition, primates such as gibbons, leaf monkeys, and the slow loris have been captured for pets. Together, these threats are driving Javan primates toward the brink of extinction. Therefore, effective conservation measures are urgently needed. There is a need to understand how habitat changes are contributing to declines in primate population density so that conservation practitioners can guide land-use practices and conservation interventions, as well as to reduce the rate of habitat change itself (Gaveau et al. 2009).

Primate watching is not as widespread as bird watching, which in the USA alone involves as many as 40-million people generating significant ecotourism business. If primate watching were to increase in popularity, it too might create significant ecotourism business opportunities. Dr. Russell A. Mittermeier first proposed the idea of primate watching, or what he called Primate life-listing. In his book "Lemurs of Madagascar", he introduced the idea of Lemur-watching and Lemur Life-listing as a new hobby that aimed to copy the success of bird-watching (Mittermeier et al. 2010). Birds can be seen everywhere including cities and gardens as well as more natural habitats. However, primates can only be seen in the wild if you go to where the primate habitat is, mostly in the tropical forests of Asia, Africa, and South America, so a lot more effort is involved in primate watching. No doubt, fewer people will become primate watchers, but the rewards are commensurate with the effort. When you make the effort and go "primate watching", make a list of what you see and note where you saw it (Mittermeier et al. 2010).

Supriatna (2019a, b) published a book entitled *Field Guide to Indonesian Primates* that describes 61 species of Indonesia's primates, their conservation status, their ecology, behavior, distribution, and where they can be observed. This book was written and inspired by many field guide books that have been written in many different countries for the express purpose of introducing tourists to primates. The goal of this paper is to review the tourism activities relating to one of the Indonesian primate species mentioned in the book. The tourism activities relating to the Javan gibbon occur in three different sites in Java: Bodogol Education Center in the Gunung Gede National Park, Gunung Halimun, Salak National Park in West Java, and Swara Owa in Central Java.

6.2 Wildlife Tourism Plus Gibbon Tourism

In the development of tourism in national parks, ecotourism has the ability to build and develop a sustainable, environmental, and society-friendly tourism (O'Brien 1999). Some aspects of nature that support the above, include:

1. Ecotourism is highly dependent on the quality of nature resources as well as historical and cultural heritage. Biodiversity is the major attraction for wildlife tourism; therefore, the quality, sustainability, and the preservation of natural resources as well as the historical and cultural heritage are very important (Hall 2010). The development of wildlife tourism also provides a significant opportunity to promote the conservation of biodiversity of Indonesia at international, national, regional, and local levels. The national park area has an enormous potential because it represents the largest unmodified contiguous habitat as well as represents a complete ecosystem.

2. Community involvement is very important because they often have a more indepth knowledge of the local natural attractions. Therefore, community involvement is a priority, starting from the planning up to the management level. For example, at the Mt. Halimun Salak National Park, over 100,000 people live within the park. Most of them settled in the national park before it was established into a national park and or when it was still a production and reserve forest. The indigenous community in Cipta Gelar is led hereditarily by a village leader of the family Abah Anom. They have long practiced forest conservation in their own way using customs that divide the forest into three zones, namely: forest Tutupan (closed forest), Titipan (deposited), and Usaha (used). Of those three zones, the most important zone is Tutupan which cannot be interfered with as it represents sacred forest. This forest is often the place where the water springs are located (Whitten et al. 1996; Harada 2005).

3. Ecotourism increases awareness and appreciation of nature, values of heritage, and culture. That is, wildlife tourism provides knowledge and experience to visitors and the local community. The value of this knowledge and experience can result in changes in the behavior of visitors, community, and tourism developers, by making them more aware and appreciative of nature and the value of historical and cultural heritage. In the United States, a love of nature is influenced by many writers who tell of the beauty of Nature such as Aldo Leopold, John Muir, and others who encouraged local communities to ensure that the government maintains the area. This led to a movement to love and appreciate nature, which eventually led to the development of established conservation areas in the United States in the eighteenth century (Primack 2002).

4. The ecotourism market at international and national levels is growing very fast. There is a trend toward increased demand for wildlife tourism products both at international and national levels. This is due to increased promotions that encourage people to behave positively toward nature and stimulate the desire to visit areas that are still in its natural conditions in order to raise awareness, appreciation, and concern for nature, plus the values of the local history/wisdom and

cultural heritage. Even some movie stars, Heads of States, and business people, travel to national parks to see exotic and charismatic wildlife. The Russian President, Vladimir Putin, became a leader who cares about tigers. A Hollywood actress visited the Tanjung Puting National Park in Central Kalimantan Borneo to see the orangutans. Recently, Harrison Ford, a movie star, became very angry because there were damages made to the forest in Tesso Nilo National Park. Awareness by the general public of the issues facing this national park led to numerous volunteers helping to reverse the damage.

5. Ecotourism serves as a means of establishing a sustainable economy. Wildlife tourism provides an opportunity for the organizers, governments, and local communities to earn a profit through nonextractive and nonconsumptive activities which enhance the local economy (Kiper 2013).

Ecotourism is not just as a special pattern of tourism activity, but it is a tourism concept, which reflects insight into the environment and follows the rules of balance and preservation of nature. The development of wildlife tourism must improve the quality of human relations, improve the quality of life for the local communities, and maintain the quality of the environment. Therefore, in its development, especially in protected areas, these considerations should be seriously taken into account (Filon et al. 1995; Sellars 1997).

Ecotourism in Java is carried out mostly at the national parks. The Javan rhino (*Rhinoceros sondaicus*) has attracted many people to see a rhino endemic to Java. Since this species is so elusive, most people only see the rhino footprint, food waste, or rhino dung. But while searching for the rhino, they also view the rain forest of Ujung Kulon National Park. In Baluran and Alas Purwo National Parks, wildlife tourism has become the main attraction of those parks. Those two national parks are comprised of savannah like habitats where grazing mammals such as deers, bantengs, wild boars, and wild dogs can be seen from an observation tower. Occasionally, some nocturnal animals such as the Javan panther (*Panthera pardus javanicus*) can be observed in early evening (Supriatna 2016).

In West and Central Java, birding is mostly common attraction, but primate watching is also recently booming. In Pangandaran nature reserve (1000 ha), there is an ecotour to see the Javan leaf monkey (*Trachypithecus margiratus*) and the long-tailed macaque (*Macaca fascicularis*). Pangandaran is a favored tourist destination in Java due to its long-white sand beach. Unfortunately, no gibbons have been reported in this area, but are found in smaller nature reserves such as Gunung Tilu Nature Reserve (3000 ha) (Supriatna 2019a, b).

6.3 Javan Gibbon Distribution

Gibbons, commonly referred to as lesser apes or small apes, are one of man's closest living relatives. They resemble each other much more closely than they resemble other mammals. This similarity is the result of a common ancestry. They look very human and often seem to behave in human ways. They play, investigate, manipulate new objects, learn fairly quickly, and communicate with each other. Some of them have been observed using tools to obtain food and even occasionally make those tools. Primates such as gibbons form complex social groups and develop behavioral patterns that are often similar to the structure of human societies (Supriatna et al. 2010; Mittermeier et al. 2013).

Javan gibbons are monogamous with a family group containing four to six individuals. A group usually consists of one adult male and one female with one or two juveniles and one to two dependent offspring. Generally, each family occupies their own territory. They average 1400 m of travel during the day. They are an arboreal species meaning they travel through the forest canopy. Gibbons are very rarely seen on the ground (Kappeler 1984; Supriatna et al. 2010). Male Javan gibbons weigh between 4.3 and 7.9 kg while females weigh between 4.1and 6.8 kg. In general, body coloration of this species varies from blackish to silvery gray to a gray-brownish color. Their face is totally black with pale eyebrows. The hair color surrounding their face is often paler to the hair color of the rest of the pelage. They have either a black or gray cap and black genital spots are present. In central Java, the gibbon coloration is a little bit paler compared to other areas (Andayani et al. 2001).

Among Javan gibbons, the female's song is sung as a duet with the male. However, when only females call, other females responded during four observed territory intrusions. Apparently, male gibbons approach intruding pairs alone while their mates sing. Consistent individual differences easily distinguish neighboring Javan gibbon females in the natural, thus compensating for the lack of a family-labeling male song in gibbons. They defend their territories with regular loud morning calls. Low levels of interactions are normally found in gibbons due to a lack of social partners. This is particularly the case for older infants and juveniles. Compared to the young of most other primates, they have no same-aged playmates because of their small group sizes. Intergroup encounters occur only once every two days and conflicts with other groups are infrequent. Fighting is rare. Female participation in border disputes is normally limited to calling. Sub-adult and adolescent males often participate in intergroup conflicts and join their parents in chasing a neighboring male (Supriatna 2016).

6.4 Gibbon Tourism in the Parks in West and Central Java

In addition to its educational value, wildlife-based tourism can also have conservation value. Through this activity, the wildlife tourism actor will realize the importance of wildlife existence. They will also realize and comprehend the threats that can endanger the wildlife's existence. Other than that, after participating in the wildlife tourism activity, they will also encounter hunting and trade activities, at that time they will be more aware and able to assist the officer in charge in handling such cases (McNeely 1994).

Wildlife tourism is also a good solution to decrease or minimize wildlife trade in Indonesia. Additionally, the tourism sector development will become the communities' profit substitution of their former main income through wildlife trading. Wildlife tourism opens up opportunities to the local communities to develop and increase their income, through local accommodation (homestays), guide skills, and handicrafts. The income from this type of tourism will enable the local communities to abandon their hunting activities. In order for wildlife tourism to succeed, it should be made certain that the funds generated from the tourists are sufficiently distributed to the local communities and the management.

Other benefits from wildlife tourism include: education and awareness of the environment, maintenance of cultural identity, and the potential opportunity to improve the economics of the local population as well as preserve wildlife and its natural environment. If managed well, wildlife tourism attractions can become sources of income for local governments and the state. For instance, Costa Rica's income from Wildlife Safari Tours accounts for almost 10% of the state's income (Fenell 1999).

Although it has the potential to deliver tangible benefits to wildlife conservation, wildlife tourism cannot simply escape from the risks of negative impacts. Wildlife tourism facilities might provide a false image and experience to the tourists. Generally, it is caused by wildlife tourist programs that do not embrace environmental and social issues. Sometimes, wildlife tourism endangers the biodiversity because tourists do not have the awareness and sense of responsibility which could damage the environment in the region of forests, lakes, sea, and beaches that are part of the conservation and wildlife tourism areas.

The potential of wildlife tourism in Indonesia is huge, especially the tourism area of sight-seeing and wildlife tours. The problem is the lack of government attention, developers, NGOs, and tourism experts in Indonesia who are seriously focusing on wildlife tourism development. Recently, there was a survey of foreign tourists visiting Indonesia and the result was amazing, where nearly 60% said they were interested because Indonesia's nature is so amazingly beautiful.

What is also very important is the desire and commitment of the Ministry of Forestry as the forest area managers and the Ministry of Maritime Affairs and Fisheries (KKP) to develop a nature marine tourism region. If the collaboration between the different expertise in the planning to the implementation of the program can be made more synergistic, adaptive, and adopt the system or the new paradigm as the program co-management, co-finance, co- responsibility, and eventually co-ownership, with nature conservation area stakeholders and local communities, then the wildlife tourism businesses will become the main tourism attraction and will be well implemented. Wildlife tourism is not only for backpackers and adventurous tourists, but also can become a sustainable tourism, which is ecology-minded, and not inexpensive.

6.5   Description of Ecotourism Sites

6.5.1   Bodogol Conservation and Education Center of the Gunung Gede Pangrango National Park

Gunung Gede Pangrango National Park (GGPNP) in West Java, Indonesia, is one of the last remaining tropical rainforests on Java. The GGPNP has a total area of 24,270 hectares and contains high biodiversity. It is the habitat of the critically endangered Javan Gibbons (*Hylobates moloch*), the Javan Hawk Eagle (*Spizaetus bartelsi*), Grizzled leaf monkey (*Presbytis comata*), Javan Leopard (*Panthera pardus melas*), and Javan slow loris (*Nycticebus javanicus*). Encroachment is the main threat to the habitat of species in GGPNP as the local people still do not have adequate knowledge and awareness on the importance of conservation (Supriatna 2006).

Tropical forests on Java, particularly mountain or sub alpine forest, have been severely degraded since the 1970s, following the rapid growth of the island's human population and the need for agriculture land and development. GGPNP is one of the few conservation areas in the region that is well preserved. Established in 1980 as one of the first national parks in Indonesia, GGPNP has been declared as one of Indonesia's six Biosphere Reserves by UNESCO.

GGPNP also is the core site of a larger significant water catchments area of 100,000 ha, which includes the neighboring mountains of Halimun and Salak. The service value of water within this region is worth ~US$100 million/year for the consumption of approximately 20 million people in 144 villages and five nearby cities, including the Capital city, Jakarta.

Only a 90-min drive from Jakarta, GGPNP is an oasis on Java, the most densely populated island in the world. Easy access by highway from three major cities, Jakarta, Bogor, and Bandung, makes this the most frequently visited national park in Indonesia. More than 50,000 visitors came to GGPNP each year, of whom 35% are students. Nevertheless, surveys conducted the park authority and a tourism institute in Bandung show the vast majority of visitors lack an understanding of their region's unique biological heritage.

Recognizing the GGPNP's ideal accessibility and existing, on 1997 a consortium was formed with Conservation International, the Agency for Forest Protection and Nature Conservation (PHPA) (Part of the ministry of Forestry with authority over GGPNP), and the Alam Mitra Indonesia Foundation (ALAMI), an Indonesian NGO whose mission is to develop and increase the participation of Indonesian people in conservation and sustainable use of the country's natural resources. At the time, the consortium agreed to develop a conservation education program in the National Park, namely Bodogol Conservation Education Center (BCEC).

The program developed includes establishing a separate entrance from the main gate to the GGNP to facilitate access by visitors interested in nature education and training. This entrance is closer to Jakarta than the main gate and avoids the traffic congestion associated with weekend visitor to the park. The facilities at this alternative gate away will promote experiential education and allow visitors to spend several days within the center. The facilities that have been established since 1998 include a 100-m canopy bridge and canopy walkway, one classroom, two furnished dormitories of 40 bunk beds each, a kitchen and restaurant, a gazebo, a park guide house, a volunteer house, display rooms, and 2 km of forest tracks with scenic outlooks that include information about the park's biodiversity, ecosystems, topography, and distance from the main gate.

BCEC is designed to be an alternative informal education that provides opportunity for target groups to explore and have immediate experience with the tropical rainforest and its surroundings. Within this BCEC, we propose to take steps to try and ensure the tropical rainforest and its biodiversity future through increase awareness and protection of its remaining ecosystems. Program aimed at increasing awareness and understanding about the importance of conserving tropical rain forest is the main issue or the principal focus (Supriatna 2006).

The program runs under the themes of "Reveals the Secret of the Rainforest". Educational contents are classified under small topic such as "Forest, the Food Supplier", "Forest, the Drug Store", and "Life in a Canopy". The Conservation Education program is designed to be suitable for different types of the visitor groups' characteristic of emphasis on providing first hand experiences on exploring the tropical rainforest. The method in delivering nature conservation issues was fun and interactive, in order to be able to encourage to curiosity and creativity, as well as positive and active participatory. The center is designed to serve the following audiences: children from surrounding urban and rural areas through the Nature Kid Program; student conservation and scientific professionals, and other interested member of the public; and Indonesia business executives over the weekend. Fees from this program would help support the maintenance and operation of the facilities.

The program, targeted primarily for students in every level, would also be designed to reach wider and prospective audiences, such as family and businessman. Also, an important target participant is member of nongovernment organization, especially in order to enhance awareness of strengthening their knowledge and capacity in nature conservation issues. Support of both the local communities, including women's participation, surrounding and living within the park, and the government, particularly those from the Ministry of Forestry and Regional Planning were taken seriously. The educational center at the park is the focus of outreach activities. Training courses for teachers, local rangers, and government officials were held at the park. Feedback from monitoring programs were used to support the park management.

In the processed program, the present education program and environmental monitoring will be continued and revised, along with addition of developing innovative education material, and capacity strengthening (through series of training for various target groups). More importantly, outreach and initiation of ecotourism program that encourage community, and especially women's, participation are going to be implemented. The 30-meter-high canopy walk is the highlight of conservation education program in the National Park, which allows visitors to appreciate rain forest from a bird's eye view. The canopy has multiple functions for recreation as well as education. It also serves as an "observation deck" for keen adventurers and research fellows to learn about animal behavior and endemic plants. From the walk, school children and other visitors also receive interpretation module on rainforest and learn how to value forest for human life. Most activities (75%) are conducted outdoors and focus on providing visitors with a first-hand experience of nature using their senses and through guided classroom activities. Other sessions (25%) are focused on introducing visitors to the forest and providing them with information about life in the forest. The outdoor activities primarily involve guided walks (including crossing the canopy bridge), nature games, and discussions. Participants are divided into small groups, usually with a maximum of six people, and are accompanied by a facilitator.

During the period from 1998 to 2013, 66.5% of the visitors to the Bodogol Conservation Education Center were students (Fig. 6.1). During this time, more than 50,000 people visited the center, including local school children, families, community groups, decision makers, and corporate executives. Overall, the hope is that students will increase their curiosity and enhance their sense of biophilia.

The Center also offers guided nature walks, lectures, and training courses. A modest research center provides laboratory space and accommodations for visiting researchers at the following costs: one-day visit (including a program + ticket + insurance + guide + welcome drink) plus the canopy bridge or the Cikaweni waterfall (3–4 h) for $3.5 (general public) or $2.5 (student) and for both the canopy bridge and the Cikaweni waterfall (4–6 h) for $5 (public) or $3 (student). The Adventure + the canopy bridge (4–6 h) cost approximately $5 (public) or $3.5

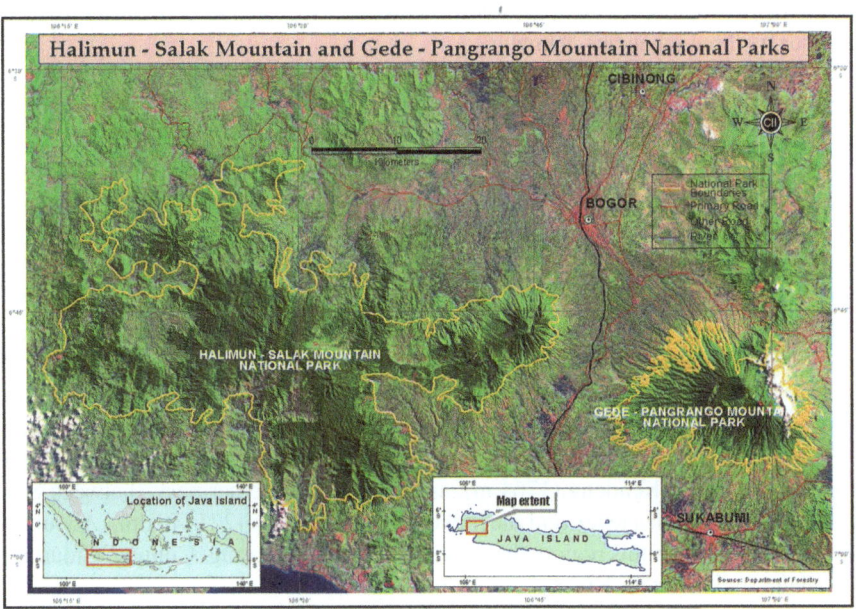

Fig. 6.1 Map of Gunung Gede Pangrango NP and Gunung Halimun-Salak NP

(student), while visiting the Cipadaranten waterfall (6–7 h for a minimum of five people) costs $6 (public) or $4 (student).

There are also several packages for multiple-day visits that include program fees, tickets, insurance, interpreters, accommodations, welcome drinks, meals, and transportation. Costs range from $20 to $50 for one person for research tourism, family gatherings, backcountry fun, family camping, and other packages. The programs also provide benefits for the communities around the park, such as employment for field staff and interpreters and income from meal provision, local motorcycle transportation (ojek), and jeep rental.

During the period from 2003 to 2007, with support from Ford Motor Company in Jakarta, the consortium launched a mobile unit that visited hundreds of schools and thousands of students surrounding the park. The car was called "Molly and Telsi" and symbolized two flagship animals; Molly is Javan gibbon (*Hylobates moloch*) and Telsi is a Javan eagle (*Nisaetus bartelsi*). The car was equipped to show documentary movies and host a talk show for school children and people in the villages. More than 40,000 people were visited by this unit.

6.5.2 Gunung Halimun National Park: Citalahab Ecotourism

The Mountain Halimun Salak National Park is located geographically between the $6^0 37'$-$6^0 53$ 'south latitude and $106^0 21'$-$106^0 38'$ east longitude, with a distance of about 100 km to the south-west of the city, 20 km to the southwest of the city of Bogor or 10 km north of Pelabuhan Ratu. This area is located in West Java Province and covers three regencies, namely:

- Bogor Regency consisting of five regencies and 13 villages.
- Sukabumi Regency consisting of three regencies and 18 villages.
- Lebak Regency consisting of four regencies and 19 villages.

The original size of the Mt. Halimun National Park is 40,000 ha, divided into 9950 ha in the Sukabumi area, an area of 14.020 ha Lebak Regency, and Bogor Regency area 16,030 ha (Fig. 6.2). Recently, additional forest included in park is the forest area of Mount Salak, which was previously given the status of protected forest/ forest reserve, so that the total width of the national park became 113,357 ha and is renamed as the Gunung Halimun Salak National Park.

The Mt. Halimun Salak National Park is a mountainous region that with more than ten mountains/hills, among them is the Mt. Halimun (1929 m), Mt. Sanggabuana (1919 m), the Mt. Halimun South (1744 m), Mt. Botol (1785 m), Mt. Amdan (1463 m), and Mt. Kendeng (1764 m), as well as several other peaks between 800 and 1200 m above sea level, and the highest is Mount Salak (2211 m dpl). About 50 rivers disgorge from this region, these rivers including Ciberang/ Ciujung, Cidurian, Cisadane, and Cimadur River.

Historically this area was recorded as the habitat of the Javan tiger (*Panthera tigris sondaica*) and the Javan rhino (*Rhinoceros sondaicus*). The biodiversity and

Fig 6.2 MT. Halimun Salak National Park Ma

ecosystems in the Mt. Halimun Salak National Park are very diverse. This area serves as a life support system, particularly its climate and hydrological functions to the Bogor, Lebak, Sukabumi Regencies and Jakarta. It also serves to support development in the surrounding region, and for the interest of science, education, and training, as well as to support the cultivation and nature tourism.

The Halimun Salak National Park has an evergreen rain forest which is the largest on the island of Java, and 20% of the total area of the lowland forests are clustered in separate plots and are mostly found around the national park. There are about 1000 species of plants including rare orchids and 17 species of Ficus. The type of ecosystem in this region can be distinguished based on height as follows: High land rain forest (500–1000 m asl) zone - in this zone, a lot of damage has been experienced and therefore it has become a secondary forest, it has a lot of wildly growing undergrowth and tree pioneers; Sub-montane forest zone (1000–1500 m above sea level) - lower montane forests have high species diversity, dominant species include the rasamala (*Altingia excelsa*), puspa (*Schima walichii*), pasang/oak (*Lithocarpus sp.*), suren (Toona sinensis), jamuju (Dacrycarpus imbricatus), baros (Magnolia blumei), waru sintok (*Cinnamomum sintok*), kiputri (*Podocarpus neriifolius*), *Antidesma montanum, Eurya acuminata, Evodia aromatica,* and various species of Fagaceae with just a low level of undergrowth; The montane forest zone (above an altitude of 1500 m above sea level) is dominated by species of the Fagaceae family such as the pasang (Quercus sp.), jamuju (*Dacrycarpus imbricatus*), and the kiputri (*Podocarpus neriifolius*).

The species of wildlife that can be found in the Halimun Salak National Park includes 11 species of squirrels, seven species of bats, seven species of otters, and five species of primates, *Cuon alpinus* (wild dog/ajag), *Manis javanica* (pangolin), *Sus scrofa* (wild boar), *Tragulus javanicus* (kanchil), and *Mydaus javanensis* (skunk). There are three species of primates endemic to Java, those are *Hylobates moloch* (Javan gibbon), *Presbytis comata* (Javan Surili), and *Trachypithecus margiratus* (Javan langur) and *Nycticebus javanicus* (Javan Coucang) also lives in the largest national park on Java Island. At this national park, the Javan gibbon population is the largest in the world, more than 1600 animals (Supriatna et al. 2003). Many different cats can be found in the park including Javan leopard (*Panthera pardus melas*), Sunda Leopard Cat (*Prionailurus javanensis*), Fishing cat (*Prionailurus viverrinus*), and Marbled cat (*Pardofelis marmorota*).

In this park, it found more than 250 species of birds (30 of them are endemic). While the endemic Javan birds that live in the national park are amongst others: Java ciung-air (Macronous flavicollis), Javan Eagle (*Spizaetus bartelsi*), Javan Quail gonggong (*Arborophila javanica*), Javan Celepuk (*Otus angelinae*), Javan Ciung mungkal (*Cochoa azurea*), Javanese Ciung air (*Macronous flavicollis*), Javanese Wergan (*Alcippe pyrrhoptera*), Javan Tesia (*Tesia superciliaris*), Javan Cerecet (*Psaltria exilis*), Javan Opior (*Lophozosterops javanicus*), and Takur tohtor (*Megalaima armillaris*).

Location for gibbon tourism is located in the Citalahab – Cikaniki area. It has an ecotourism center inside the tropical rainforest there and primates are known to occur there, e.g., Javan Surili (*Presbytis comata*), Javan Gibbon (*Hylobates moloch*), Javan Langur (*Trachypithecus margiratus*), Long-tailed Macaque (*Macaca fascicularis*), and Javang councang (*Nycticebus javanicus*) (Supriatna 2014, Supriatna 2019). Estimates of the number of Javan gibbons in Halimun-Salak vary, but range between 900 and 1220 individuals (Kool 1992; Asquith et al. 1995; Sugardjito and Sinaga 1999; Nijman 2015), and it is estimated that 330–400 km² of suitable habitat remains for the gibbons (Rinaldi 2003; Djanubudiman et al. 2004; Nijman 2004).

This gibbon tourism is a community-based effort. Originally, it was created by Gunung Gede Pangrango Halimun Salak consortium. A consortium of 30 stakeholders included national parks, local governments, research institutions, universities, NGOs, and private sectors' collaboration to develop the Gunung Gede Pangrango and Halimun Salak national parks. Then, the community took the initiative to develop ecotourism in the Citalahab site. There are also several attractions close by the Citalahab village. Camping sites can be found in the village Cikaniki and Citalahab, located at the road between the Nirmala Tea Plantation and Kabandungan village. The Canopy Trail has a length of 110 m and is located near in the village Cikaniki just behind the research station of the park authority. The supporting facilities present in the Mt. Halimun National Park are amongst others, the Pondok Kerja (work cabin), Watch Posts, Guest Inn, Information Center, and Hiking trails.

In this park, the communities residing in the surrounding area consist mostly of the Sundanese tribe, especially from the Kasepuhan Citorek and Cicemet community, who still adhere to or are still following their cultural traditions. In this society, there are several species of traditional ceremonies, among others:

- Nandur, is done when harvesting of paddy is about to start.
- Meupeuk pare berkah, is done when the paddy is starting to bear rice.
- Nganyaaran, is done when they are going to store the harvested paddy into designated rice/ paddy storage sheds/ barns.
- Seren Tahun, is done by the Kasepuhan Banten Kidul community around the month July, as a sign that the farming period over the past year is over (Thanksgiving).
- Ngaruwah, thanksgiving for the Ruwah month.

The Kasepuhan community are also residing in the village Cicarucub, Cisungsang, Bayah, and South Banten/Kidul. In the western region you will find the Badui tribe settlement (indigenous people of West Java), who still uphold their traditional way of life and it can be said that it has not been influenced by other cultures. In the surroundings of the region there are 44 villages, a small part of them are in the buffer zone of the National Park, and four villages are located in the region's enclave.

The Kasepuhan community has a unique pattern of forest management, which is similar to the zoning concept in the pattern of modern conservation. The forest is divided into four zones, namely Leuweung Kolot (not to be disturbed), Leuweung Titipan (must have permission from the Girang Elders/traditional leaders), Leuweung Sirah Cai (forest as source of water), and Leuweung bukaan (can be used) (Harada 2005). The utilization of forest products by the communities in the form of construction timber and household appliances, firewood, ferns, ornamental plants, rattan, plant food, medicinal plants, and herbs needed for supplies needed in traditional ceremonies is quite high.

To visit this center, you can travel by car from Bogor to Leuwiliang, which is 20 km and takes about 30 min on public transport and from Leuwiliang to Nanggung is a further 15 km and takes about 20 min on public transport. From Nanggung to Cisangku is another 15 km from there and can be reached with a motor bike or car ride of an hour. The site can also be reached using different routes, such as from Jakarta to Parungkuda (80 km), first using bus then taking public transport from Parungkuda to Cipeuteuy (30 km). From Cipeuteuy, there is no public transport, so it requires renting a motor bike to Perkebunan Teh Nirmala (Citalahab) for 2–3 h.

6.5.3 Gibbon Coffee: Gibbon Watching and Shade Grown Coffee Eco-Tour

Comprehensive survey on Javan gibbon (*Hylobates moloch*) in Java has been initiated since eighties, the survey on this species has been conducted by Kappeler (1984). He has assessed 32 forest patches that are inhabited by Javan gibbon in Java and the population has been estimated between 2400 and 7900 individuals. Sugardjito and Sinaga (1999) have studied the population specifically, in Halimun National Park, West Java and have been estimated the population about 1000 individuals. Further survey, conservation activities, and ecological study of Javan gibbon also continued intensively in this area (Iskandar 2007; Kim et al. 2011). Based on the published data available on gibbon population, Supriatna (2006) estimated that some 2000–4000 gibbons remaining in the wild. The latest survey of Javan gibbon was conducted by Nijman and van Balen (1998, b) in Dieng Mountains with the results of remaining population between 519 and 577 individuals, recent survey of Javan gibbon conducted by Djanubudiman et al. (2004), visited 23 locations in West Java and seven locations in Central Java, has provided population estimate of gibbons 492 individuals in Dieng Mountains and 96 individuals gibbons in Mount Slamet and the total estimate for the whole of Java 4888 individuals. Nijman (2004) also stated that the total gibbon population in Java is between 4000 and 4500 individuals. In contrast to West Java, the forest habitat in Central Java does not have any protected area which forms a network system and received little attention among conservationists and researchers. Consequently, forest habitat in this region is more threatened by encroachment. However, based on the field survey conducted by previous authors, there still remained a large forest block in Central Java where the Javan gibbon lived.

Setiawan et al. (2012) conducted population and distribution survey of Javan gibbon in Central Java, visiting all locations that have reported on the previous reports. There are 16 locations in Central Java occupied by the gibbons that clumped but fragmented in two largest forest blocks in Dieng Mountains and Mt. Slamet and from the result they found that there are 51 gibbons (21 groups) in this forest with the density 7.57 individuals/km^2, this number is highest density in Central Java, where there is no conservation area that protecting gibbon habitat. Threats for gibbon population and habitat are different among locations and this could be used for determining priorities in conservation management for this species (Nijman 2004; Supriatna 2006; Supriatna et al. 2010). Habitat degradations due to human activities in the forest for logging, encroachment, and hunting are major threats. Expansion of coffee plantation was encroaching to the forest, especially in the lowland area where gibbon has suitable for food resources and climates. The lowland area is also a good habitat for various commodities for source of income of communities.

The project focused in Sokokembang village, created by a group of primatologists and conservationists founded an organization called SWARAOWA (https://swaraowa.org/). This group managed the project activities included tourism and coffee development. The project's goal is to conserve Javan gibbon in their native habitat in Petungkriyono District, Western part of Dieng Mountain, as the eastern most range distribution of Javan gibbon. This organization's aim is to create sustainable economy initiative to preserve Javan gibbon and enhance economy income of local community.

The Coffee and Primate Conservation Project was initiated in 2012 in Sokokembang forest. Sokokembang is located at Petungkriyono District, Pekalongan Regency, Central Java, Indonesia. Pekalongan could be reached about 7 h from Jakarta by train, from Semarang International airport about 2 h and about 6 h from

Yogyakarta. The forested area consists of mountainous region started from 450 meter up to 1400 meters above sea level, about 45 min by car from the Pekalongan city center of Central Java. To find Sokokembang, just follow the main road from Pekalongan to Petungkriyono, through Doro, then the forest started from valley of Welo river, only one road through the forest and this is the most active road at the moment for tourism activities in upper area in Petungkriyono district.

The gibbons can be found along the road from Kroyakan to Sokokembang village, approximately 6 km long narrow and winding road. The forest canopy is densely found just typically rainforest, where you can see the gibbon swinging on trees. Coffee was cultivated here during the colonial era and most coffee plants are grown on government land. As a result, the community can harvest the coffee cherries even if they do not own the land. Without any proper management, coffee was continually planted in the javan gibbon forest which involved chopping down the native trees for more coffee plants. Coffee therefore became one of the threats for habitat destruction without any additional value for community surrounding the forest.

Then, the project realized that coffee as commodities is overlapped interest among human, gibbons, and forest. Quick cash is needed for all famers here, so sometime they get goods from the trader in the market, and when they have coffee, the farmers bring to them. The forest as javan gibbon habitat will also degrade due to expansion of coffee plants, no economic value for the communities compares to cost production and ecological impact.

Coffee and Primate conservation project introduced to the community how to grow shade grown coffee and enhance the value of coffee through postharvesting management. The project thought modern processing techniques and marketing strategies to sell local coffee beans to larger market at higher prices. It is a long story of a successful project, where coffee culture actually did not exist in this region. However, since 2012, the project was initiated, and now this new culture has been developed. Kopi Owa was born in the local community, in the Javan gibbon habitat of Sokokembang, as one of the commodities that help to protect the gibbon and community. The Kopi Owa in local language has been become identity and sustainable enterprise that support conservation of Javan gibbon and economy support for the community. Through the Kopi Owa network, the project has created a farmers network, who previously worked as wildlife hunters, to produce the gibbon coffee, reduce pressure on forest habitats, prevent the chopping down of trees, and properly manage shade grown coffee and better coffee processing and marketing. Shaded trees are important for the gibbon but also better for coffee taste too.

Sokokembang village is the easiest location to watch the gibbon in Central Java, where every morning the gibbon song can be heard clearly from the village. Along the road from Sokokembang- Kroyakan have become regular activities for Javan gibbon monitoring, given the higher possibility to see and use this road in the forest for conservation education about Javan gibbon and other wildlife. Shade grown coffee found in the forest nearby the village is also accessible for wildlife enthusiasts who are looking for different experience.

Coffee and Gibbon are perfect blend created in this project and community in Sokokembang. Increasing nature tourism activities in this area, Sokokembang and its community assisted by SWARAOWA's team provide alternative ecofriendly tour. Gibbon watching conducted using doplak (a local pick-up vehicle) that modified for people standing in the back becomes different experience to watch the gibbon without exhausted to walk way. This trip normally conducted at least three people at one day finish, with special guide from local community or SWARAOwa's team. Gibbon spotting started at 5.30 a.m. and will be finished at 9.30 a.m. The guide will introduce not only the gibbon behaviors, important value, and cultural perception but also other primate sightings, such Javan langur (*Trachypithecus auratus*), Javan Surili (*Presbytis comata*), and Long-tailed macaque (*Macaca fascicularis*).

The trip is then continued with coffee adventure to see how farmers manage shade grown coffee, starting from the shade forest to the cup. During the coffee tour, the visitor is guided by the SWARAOWA team, who will introduce the importance of agroforest coffee habitat, both for biodiversity and the economy. Coffee, in particular, has the potential to benefit both livelihoods and the environment. Coffee, like cacao, is an understory shrub or tree that thrives under the shade of diverse forest trees. Some of the species have important roles for seed dispersers, they help forest regeneration, they require no fertilizer when growing in this forest habitat, and they have longer times for mature fruits, meaning it will result in better beans. Through this coffee adventure, guests will be introduced to the agroforest area that produces various commodities at various times, which is very important family income and generally for food security.

Coffee harvesting and its processing is only available during coffee season, usually July–August for robusta coffee. Basic coffee roasting will be introduced to guests, and coffee cupping is also an option to learn about various coffee tastes and aromas. Coffee packaging with gibbon stories will be attached on the coffee package so that everyone can take home a good souvenir.

References

Andayani N, Morales JC, Forstner MRJ, Supriatna J, Melnick DM (2001) Genetic variability in mtDNA of the silvery gibbon: implications for the conservation of a critically endangered species. Conserv Biol 15(3):770–775

Asquith NM, Martarinza M, Sinaga RM (1995) The Javan gibbon (Hylobates moloch): status and conservation recommendations. Tropical Biodiv 3(1):1–14

Djanubudiman G, Arisona J, Iqbal M, Wibisono F, Mulcahy G, IndrawanM HRM (2004) Current distribution and conservation priorities for the Javan gibbon (*Hylobates moloch*). In: Report to great ape conservation fund, US fish and wildlife service. Indonesian Foundation for Advance of Biological Sciences and Center for Biodiversity and Conservation Studies of University of Indonesia, Depok, Washington, DC

Fenell DA (1999) Ecotourism: an introduction. Routlidge, London

Filon FL, Jacqueamot A, Reid R (1995) The important of wildlife to Canadians. Canada Wildlife Service, Canada

Gaveau DLA, Wich S, Epting J, Juhn D, Kanninen M, Leader-Williams N (2009) The future of forests and orangutans (*Pongo abelii*) in Sumatra: predicting impacts of oil palm plantations, road construction, and mechanisms for reducing carbon emissions from deforestation. Environ Res Letters 4:034013. https://doi.org/10.1088/1748-9326/4/3/034013

Hall MC (2010) Tourism and biodiversity: more significant than climate change? J Heritage Tourism 5:253–266. https://doi.org/10.080/1753873X2010.517843

Harada K (2005) Local use of agricultural lands and natural resources as the commons in Gunung Halimun National Park, West Java, Indonesia. IntJ Sust Dev & Wild Ecol 15:34–47. https://doi.org/10.1080/13504500509469616

Iskandar E (2007) Habitat dan populasi owa jawa (*Hylobates moloch* Audebert, 1797) di Taman Nasional Gunung Halimun-Salak Jawa Barat. Thesis Doktor, Institute Pertanian Bogor, Bogor, p 141

Kappeler M (1984) The gibbon in Java. In: Preuschoft H, Chivers DJ, Cheney DC, Seyfarth RM, Wrangham PW, Strushaker TT (eds) The lesser apes: evolutionary and behavioral biology. University of Chicago Press, Chicago, pp 19–31

Kim S, Lappan S, Choe JC (2011) Diet and ranging behavior of the endangered Javan gibbon (*Hylobates moloch*) in a submontane tropical rainforest. Am J Primatol 73:270–280

Kiper T (2013) The role of ecotourism in sustainable development. Chapter 31, pp 773–802. https://doi.org/10.5772/55749

McNeely J (1994) Protected areas for the 21st century: working to provide benefits to society. Biodivers Conserv 3(5):390–405. https://doi.org/10.1007/BF.00057797

Miettinen J, Shi C, Liew SC (2011) Deforestation rates in insular Southeast Asia between 2000 and 2010. Glob Chang Biol 17(7):2261e2270

Mittermeier RA, Louis EE Jr, Richardson M, Schitzer C, Langrand O, Rylands AB, Hawkins F, Rajaobelina S, Ratsimbazafy J, Nash SD (2010) Lemurs of Madagascar: tropical field guide series. Conservation International Field Guide Series, Washington, D.C.

Mittermeier RA, Rylands AB, Wilson DE (eds) (2013) Handbook of the mammals of the world, vol. 3. Primates. Lynx Editions, Barcelona

Nijman V (2004) Conservation of the Javan gibbon Hylobates moloch: population estimates, local extinctions, and conservation priorities. Raffles Bull Zoo 52(1):271–280

Nijman V, van Balen B (1998) A faunal survey of the Dieng Mountains, Central Java, Indonesia: status and distribution of endemic primate taxa. Oryx 32:145–146

O'Brien BR (1999) Our National Parks and the search for sustainability. Texas Press, Austin, p 246

Primack RB (2002) Essentials of conservation biology, 3rd edn. Sinauer Associates, Sunderland M.A

Rinaldi D (2003) The study of Javan gibbon (*Hylobates moloch* Audebert) in Gunung Halimun National Park (distribution, population and behavior). Research and Conservation of Biodiversity in Indonesia, pp 30–48

Roos C, Boonratana R, Supriatna J, Fellowes JR, Groves CP, Stephen D (2014) An updated taxonomy and conservation status review of Asian primates. Asian Primate J 4(1):2–38

Sellars RW (1997) Reviewing nature in the National Parks: a history. Yale University Press, New Heaven, p 379

Setiawan A, Nugroho TS, Wibisono Y, Ikawati V, Sugardjito J (2012) Population density and distribution of Javan gibbon (Hylobates moloch) in Central Java. Indonesia Biodiversit J Biolog Diversity 13(1)

Sugardjito J, Sinaga MH (1999) Conservation status and population distribution of primates in Gunung Halimun National Park, West Java, Indonesia. In: Proceedings of the international workshop rescue and rehabilitation. Taman Safari and University of Indonesia, Cisarua, Bogor

Supriatna J (2006) Conservation programs for endangered Javan gibbon (Hylobates moloch). Primate Conservat 21:155–162

Supriatna J (2016) Berwisata Alam di Taman Nasional (tourism at the National Park). Pustaka Obor, Jakarta

Supriatna J (2019a) Field guides to Indonesia primates. Pustaka Obor, Jakarta

Supriatna J (2019b) Field guide of Indonesia primates. Pustaka Obor, Jakarta

Supriatna J, Tilson RL, Gurmaya KJ, Manansang J, Wardojo W, Sriyanto A, Teare A, Castle K, Seal US (1994) Javan gibbon and Javan langur: population and habitat viability analysis report. IUCN/SSC Conservation Breeding Specialist Group (CBSG), Apple Valley, Minnesota, p 112

Supriatna J, Mootnick A, Andayani N (2010) In: Gursky-Doyen S, Supriatna J (eds) Javan gibbon (Hylobates moloch): population and conservation. Indonesian Primates. Springer, New York

Whitten T, Soeriaatmadja RE, Affif SA (1996) The ecology of Java and Bali. Periplus, Hongkong

Chapter 7
Encountering Sulawesi's Endemic Primates: Considerations for Developing Primate Tourism in South Sulawesi, Indonesia

Katherine T. Hanson, Kristen S. Morrow, Putu Oka Ngakan, Joshua S. Trinidad, Alison A. Zak, and Erin P. Riley

Abstract The island of Sulawesi, Indonesia is renowned as a birder and diver's paradise, attracting tourists from around the globe who seek to encounter rare bird species or abundant and unusual marine life. In contrast to other areas of Indonesia (e.g., Bali and Kalimantan), Sulawesi is less known for its primate tourism opportunities, despite being home to at least 14 endemic primate species. In this chapter, we explore the possibilities and requisite considerations for developing primate tourism in South Sulawesi, a region of the island with minimal established tourism infrastructure. We argue that cautious, thoughtful, and collaborative development of primate tourism in South Sulawesi have the potential to raise awareness of local primate biodiversity and conservation issues, supplement and diversify local livelihoods, curb the acceleration of extractive industries, and provide a valuable contrast to other primate tourism sites across Indonesia. Though the aim of this chapter is to open a dialogue among local stakeholders and international practitioners regarding responsible development of primate tourism in South Sulawesi specifically, the considerations raised here are relevant in other regions where formal primate tourism remains underdeveloped. In particular, we encourage the consideration of existing

K. T. Hanson (✉)
Department of Anthropology, University of Texas at San Antonio, San Antonio, TX, USA

K. S. Morrow
Department of Anthropology and Center for Integrative Conservation Research, University of Georgia, Athens, GA, USA

P. O. Ngakan
Faculty of Forestry, Universitas Hasanuddin, Makassar, Indonesia

J. S. Trinidad · E. P. Riley
Department of Anthropology, San Diego State University, San Diego, CA, USA

A. A. Zak
Human-Beaver Coexistence Fund, Oakton, VA, USA

© The Author(s), under exclusive license to Springer Nature Switzerland AG 2022   111
S. L. Gursky et al. (eds.), *Ecotourism and Indonesia's Primates*, Developments in Primatology: Progress and Prospects, https://doi.org/10.1007/978-3-031-14919-1_7

dimensions of human-nonhuman primate coexistence (including conflict), tourism audiences, and the degree of local engagement from diverse stakeholders.

7.1 Introduction

Sulawesi, an island best known among tourists for its unparalleled birding opportunities, picturesque diving locales, and rich cultural heritage, is not typically recognized as a primate tourism destination. In contrast to other areas of Indonesia, such as Bali or Kalimantan, primate tourism on Sulawesi is not well-developed. This is surprising given that Sulawesi is home to a number of endemic primate species, including seven macaque species (*Macaca*) and at least seven tarsier species (*Tarsius*). Additionally, it is regarded as a global biodiversity "hotspot," garnering international attention and conservation protections (Lowe 2006; IUCN 2008; Riley 2010; Shekelle et al. 2017). With notable exceptions by researchers working in North Sulawesi, very little has been written about primate tourism on Sulawesi (Kinnaird and O'Brien 1996; Melfi 2010). This chapter aims to address that gap, with a particular focus on primate tourism in South Sulawesi, Indonesia. We begin by "setting the stage" for understanding Sulawesi's tourism potential by reviewing the ecological and cultural diversity of Sulawesi. We follow this section with a brief background on tourism in Sulawesi before describing the sites where tourists and primates interface in this region, examining primate tourism in South Sulawesi as a complement to already popular nature-based tourism on the island. We then review the major factors that need to be considered in the development and management of primate tourism in South Sulawesi. These include the potential conservation benefits, the relevant ethical dimensions (e.g., ecological, biological, and behavioral impacts as well as the effects on local communities), and emerging concerns, such as the role of social media in advancing primate tourism and the implications of primate tourism in the COVID-19 era and beyond. Our objective in this chapter is to open a dialogue among local community members, protected area managers and staff, conservation practitioners, primatologists, and other researchers regarding existing patterns of interaction between tourists and primates and responsible and sustainable development of primate tourism in South Sulawesi.

7.2 Setting the Stage: *Ecological and Cultural Diversity of Sulawesi*

Sulawesi, the fourth largest island in Indonesia and the eleventh largest in the world, is both culturally and ecologically diverse, thereby making it a prime location for tourism. While analyses of rock art in the limestone karst region in Maros, South

Sulawesi suggest that humans were living on the island at least as early as 40,000 ya (Aubert et al. 2014), more recent archeological evidence (e.g., stone artifacts associated with megafaunal fossil remains) indicate that hominins may have existed on the island prior to the expansion of modern humans into Southeast Asia approximately 118,000 ya (van den Bergh et al. 2016). The current human population of Sulawesi is estimated at 19,934,000 (2020 projected estimate, Badan Pusat Statistik 2014) and comprises multiple ethnic groups (e.g., Bugis, Makassar, Mandar, Toraja, Duri, Amma Towa, Butonese, Tolaki, Kaili, Pamona, Minahasa, Sangirese, Gorontalo, Bolaang-Mongondow (Babock 1982)). Given this ethnic complexity, religion and subsistence style are typically the predominant criteria used for ethnic self-identification, either aligning with or overriding region and language as markers (Davis 1976). In addition to these major ethnic groups, Sulawesi is home to immigrants from China and Saudi Arabia, as well as transmigrants from other areas of Indonesia, such as Java and Bali. Sulawesi's linguistic diversity is also comparatively high: it is estimated that 114 native languages are spoken, all of which belong to the Malayo-Polynesian branch of the Austronesian language family (Lewis 2009). While traditional forms of subsistence include swidden (or slash and burn) agriculture and fishing (Davis 1976), today, many communities practice wet-rice agriculture and plantation agriculture of cash crops, including coffee (*Coffea* spp.), cacao *(Theobroma cacao)*, palm oil *(Elaeis guineensis)*, candlenut *(Aleurites moluccana)*, and cloves *(Syzygium aromaticum)*.

Sulawesi's ecological diversity stems from its position within Wallacea—a unique biogeographical zone that is characterized by a mix of Asian and Australasian flora and fauna (e.g., primates and marsupials) and that exhibits a remarkably high level of endemism. Of the 332 extant bird species on Sulawesi, 27% are endemic (Whitten et al. 2002), including the Sulawesi dwarf hornbill (*Rhabdotorrhinus exarhatus)* and the maleo (*Macrocephalon maleo*) (Birdlife International 2020). The level of endemism is even greater among mammals: 62% of the mammals found in Sulawesi are endemic, and that percentage rises to 98% if bats are excluded (Whitten et al. 2002). Among these endemic mammals are the enigmatic yet elusive babirusa (*Babyrousa celebensis)* and the dwarf buffalo, or *anoa (Bubalus depressicornis and B. quarlesi)*. The nonhuman primates of Sulawesi include members from two genera: *Macaca* and *Tarsius* (Figs. 7.1 and 7.2). Fooden (1969) classified the Sulawesi macaques as seven species (*Macaca nigra, M. maura, M. tonkeana, M. hecki, M. ochreata, M. brunnescens*, and *M. nigrescens*), which represent 30% of the genus in only 2% of its geographical range. The Eastern tarsier group, represented by *Tarsius*, is considered the most species-rich ($\geq$ 16 taxa) of the three clades, with at least 12 species (*T. tarsier, T. fuscus, T. sangirensis, T. dentatus, T. pumilus, T. pelengensis, T. lariang, T. tumpara, T. wallacei, T. spectrumgurskyae, T. supriatnai,* and *T. niemitzi*), but possibly more, being endemic to mainland Sulawesi (Groves and Shekelle 2010; Shekelle et al. 2019).

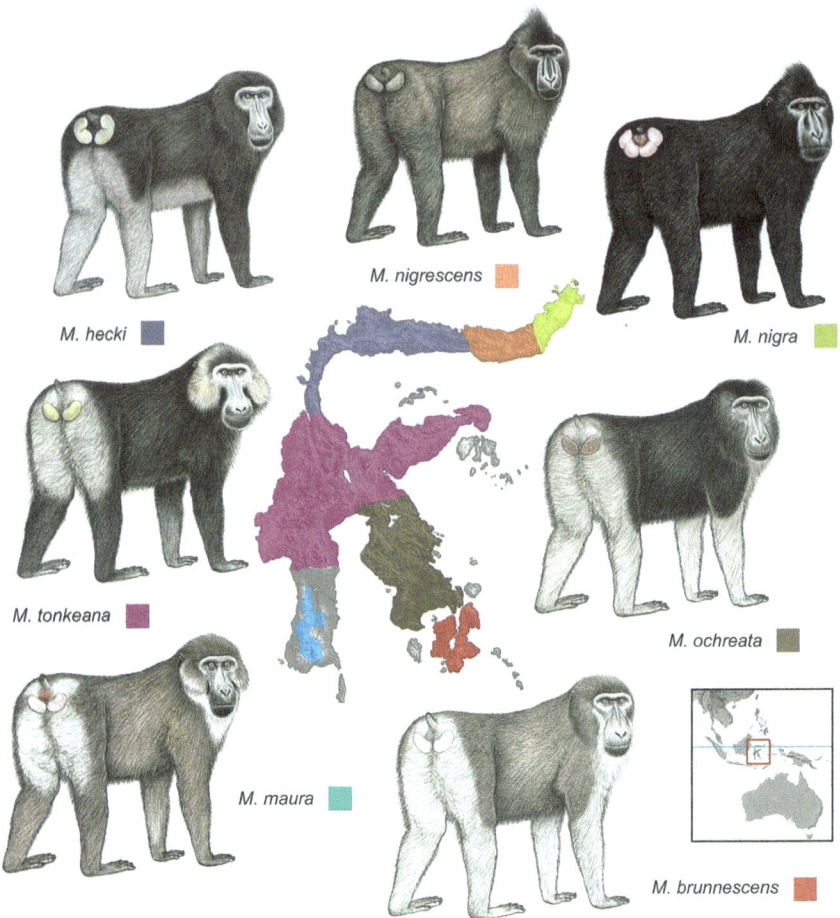

Fig. 7.1 Sulawesi's endemic macaque species (*Macaca*). (Illustration courtesy of Stephen Nash)

7.3   Tourism in Sulawesi

7.3.1   Foreign Tourism

Foreign tourism plays a significant role in Indonesia's economy. In 2018 alone, 15.81 million tourists visited the country, resulting in 16.4 billion USD in foreign exchange (Badan Pusat Statistik Indonesia 2018a; b). In response to this economic contribution, the federal government has long encouraged the expansion of tourism sites and the development of additional tourist facilities (e.g., Adams 1997; Prodjo 2017). Most foreign tourists travel to Bali, leading to an unequal

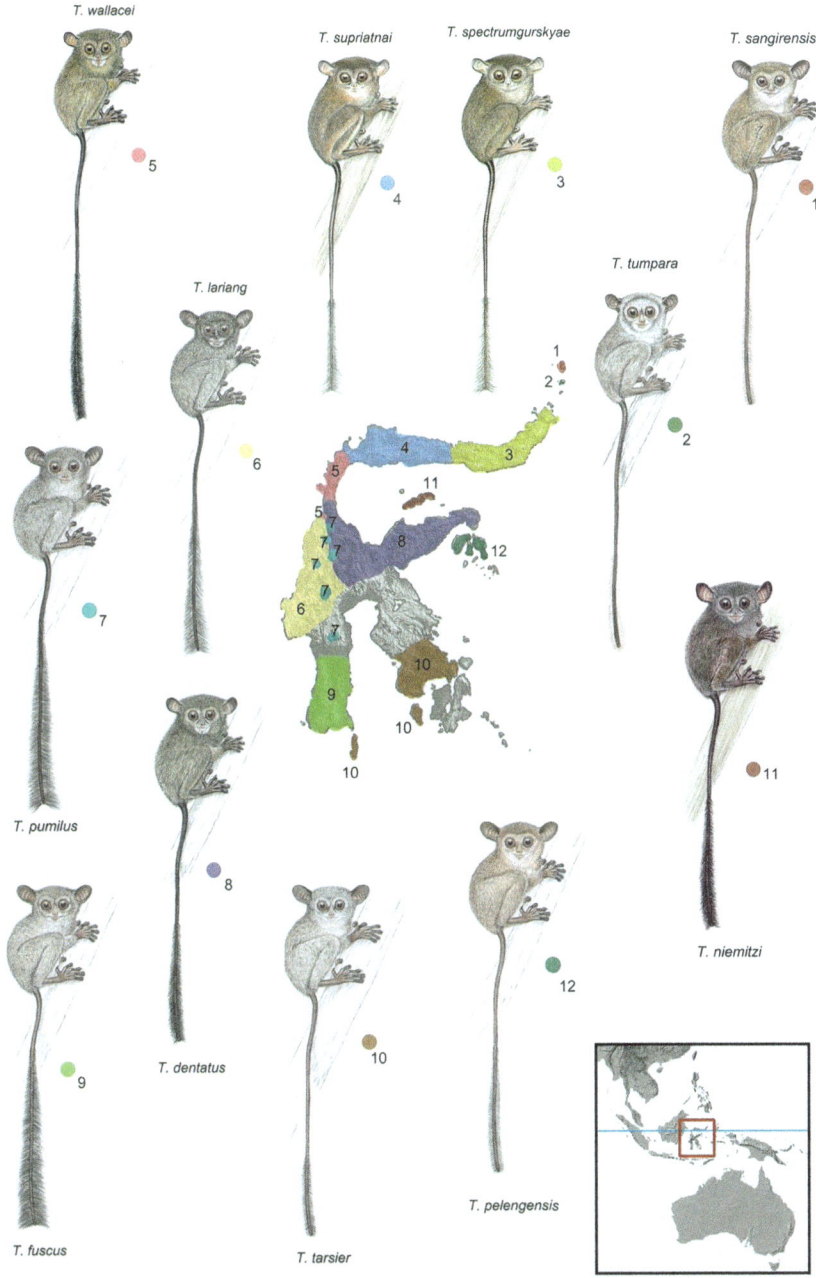

Fig. 7.2 Sulawesi's endemic tarsier species (*Tarsius*). (Illustration courtesy of Stephen Nash)

distribution of revenue and creating an unsustainable strain on Bali's resources and infrastructure (Badan Pusat Statistik 2020; Chong 2020). Sulawesi, in particular, receives a nominal number of foreign tourists who visit the country. Foreign tourism to Indonesia declined dramatically following the 1998 economic crisis and the 2002 Bali bombing, and in Sulawesi has recovered quite slowly (Junaid 2014; Pambudi et al. 2009). Although the number of tourists visiting Sulawesi increases each year, the proportion of foreign tourists remains less than 1% of the total who visit Indonesia (Table 7.1). Within Sulawesi, a far greater number of tourists arrive through North Sulawesi than South Sulawesi; in 2019 North Sulawesi received over 153,000 tourists compared to South Sulawesi's 17,771 tourists (Table 7.1; see Figure 7.3 for province designations).

7.3.2 Cultural Tourism

Despite its unique ecology and location within the Wallacea region, the most well-known tourist sites in Sulawesi focus on cultural tourism rather than nature or wildlife based tourism (Junaid 2014). The vast cultural diversity in Sulawesi offers potential resources to expand the tourism industry on this island, and regional government officials continue to actively promote tourism development (Junaid 2014; Suriamihardja 2010). The main site of cultural tourism in Sulawesi is in the Tana Toraja regency, home to the Toraja ethnic group (Junaid 2014). Tourism in Tana Toraja began in the 1970s when the Suharto administration identified it as an Outer Island destination which should be promoted to expand the tourism industry (Adams 1997). Marketing Tana Toraja to international tourists created a popular destination for witnessing novel funeral rites, visiting burial cliffs, observing traditional architecture, and viewing mountainous scenery; it was through this intentional marketing effort that Sulawesi became known as one of Indonesia's tourist destinations (Hasyim 2019; Scarduelli 2005; Yamashita 1994).

Beyond Tana Toraja, there are few other sites of cultural tourism in Sulawesi. The Bada and Besoa valleys in Lore Lindu National Park in Central Sulawesi enable

Table 7.1 Foreign arrivals through two ports of entry on the island of Sulawesi (Badan Pusat Statistik 2020)

| | Port of entry | | | | |
| | North Sulawesi | | South Sulawesi | | |
Year	Persons	% of total	Persons	% of total	Total tourists arriving to Indonesia (Persons)
2015	27,059	0.27	13,091	0.13	10,230,000
2016	47,103	0.41	16,862	0.15	11,520,000
2017	87,976	0.63	18,355	0.13	14,040,000
2018	127,879	0.81	14,126	0.09	15,810,000
2019	153,658	0.95	17,771	0.11	16.110,000

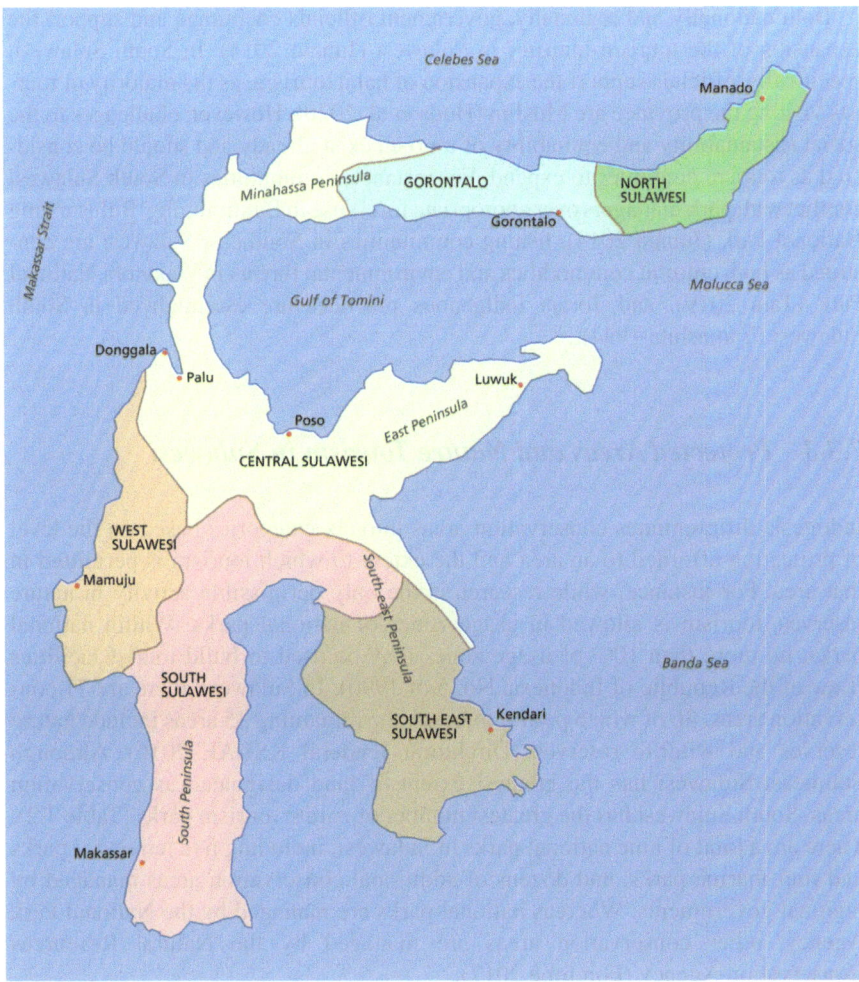

Fig. 7.3 Sulawesi's provinces featuring capital cities and surrounding bodies of water. Map from Wikimedia Commons, distributed under a CC BY-SA 3.0 license

visitors to view impressive megalith statues and cylindrical stone vats (Rahmat et al. 2016). Two villages in the Bulukumba regency in South Sulawesi offer opportunities to experience Bugis culture: in Tana Beru, tourists can observe traditional boat making processes and in Tana Toa, tourists can observe traditional houses and indigenous practices (Junaid 2014). The Somba Opu Fort in Makassar, a relic of the Gowa Kingdom, includes replicas of traditional houses of the Bugis, Makassar, Toraja, and Mandar people (Junaid 2014). Additional museums that offer cultural tourism opportunities include the Balla Lompoa Museum in Sungguminasa and the La Galigo Museum, Makassar City Museum, and historic Fort Rotterdam in Makassar (Junaid 2014).

Both nationally and regionally, government officials encourage and support the expansion of the tourism industry in Sulawesi (Junaid 2014). In South Sulawesi, specifically, officials support the expansion of halal tourism, as the majority of tourists visiting the province are Muslim (Huda et al. 2020). However, challenges to the social sustainability and equitability of tourism exist already and should be considered as tourism continues to expand. For instance, communities in South Sulawesi conflict with park managers over appropriate forest use in Bantimurung-Bulusaraung National Park (Junaid 2014), fishing communities in Southeast Sulawesi are construed as both cultural commodities and environmental threats in Wakatobi National Park (Tam 2019), and Toraja indigenous practices are essentialized in South Sulawesi (Yamashita 1994).

7.3.3   Protected Areas and Nature Tourism in Sulawesi

Indonesia differentiates conservation areas into six categories based on the level of protection afforded to an area and the extent to which tourism is permitted in that area. For instance, while research is the only permissible activity in nature reserves, tourism is allowed in usage zones of national parks. Within national parks, no more than 10% of usage zones may be used to build tourist facilities (Law of the Republic of Indonesia No. 5 of 1990). In Sulawesi, there are 71 conservation areas, 36 of which permit tourism; the remaining 35 areas include nature reserves and wildlife reserves (Direktorat Jenderal KSDAE 2016). Although Southeast Sulawesi has the greatest extent of land designated as conservation areas, South Sulawesi has the greatest number of nature tourism parks (Table 7.2). There are a total of nine national parks in Sulawesi, including five terrestrial parks and four marine parks, and dozens of additional conservation areas managed by regional governments. Whereas national parks are managed by the National Park Agency, other conservation areas are managed by the Natural Resources Conservation Agency (Forclime 2017).

Nature tourism in Sulawesi focuses primarily on marine parks, beaches, and seascapes, and is mostly located in North Sulawesi (Hakim et al. 2012). Popular marine destinations include Wakatobi National Park in Southeast Sulawesi, Bunaken Islands in North Sulawesi, Togean Islands in Central Sulawesi, and Losari Beach, Spermonde Islands, Takabonerate Islands, and the Bira Cape in South Sulawesi (Junaid 2014; Ross and Wall 1999). The most visited marine park in Sulawesi is Bunaken National Park, likely due to its proximity to Manado, the capital city of North Sulawesi (DeVantier and Turak 2004; Tangian et al. 2015). Although marine tourism is more popular, expansion of tourism capacity on Sulawesi's coastal islands is constrained by the availability of freshwater (Sahabuddin 2019; Smith 2012). One site in particular, Wakatobi National Park, has been targeted by the federal government as part of the "Ten New Bali"

Table 7.2 The number and distribution of conservation areas designated by the central government of Indonesia on Sulawesi (Forclime 2017). See Figure 3 for province designations

Province	Number (unit)						Total area (ha)
	CA	SM	TN	TWA	THR	TB	
North Sulawesi	3	2	1	3	1	0	136,210.55
Gorontalo	4	1	1	0	0	0	375,123.68
Central Sulawesi	7	6	2	4	1	1	991,013.01
West Sulawesi	0	0	1	1	0	0	214,950.35
South Sulawesi	3	1	2	8	2	1	810,978.65
Southeast Sulawesi	3	5	2	4	1	0	1,725,023.10
Total	20	15	9	20	5	2	4.253.299,34

CA = Cagar Alam (Nature Reserve), SM = Suaka Margasatwa (Wildlife Conservation Area), TN = Taman Nasional (National Park), TWA = Taman Wisata Alam (Nature Tourism Park), THR = Taman Hutan Raya (Raya Forest Park), TB = Taman Buru (Hunting Park)

program (Prodjo 2017). Launched in 2016, this program seeks to develop new tourist destinations throughout Indonesia that replicate the economic effects of tourism in Bali. The main tourist attraction in Wakatobi is diving, but beaches, local cuisine, and annual cultural festivals are also popular tourist attractions. Foreign arrivals to Wakatobi consistently increased from 2015 to 2017, reaching a total of 2904 foreign tourists in 2017; meanwhile, foreign tourists visiting Indonesia in 2017 numbered 14,040,000 (Badan Pusat Statistik 2020). The growth of Wakatobi's tourism industry has been facilitated by tourist-oriented narratives of biodiversity conservation, ecological sustainability, and economic security for local communities (Tam 2019; von Heland and Clifton 2015).

In South Sulawesi, terrestrial sites with waterfalls and caves are popular among domestic tourists. Destinations such as Malino and the Maros karst areas attract tourists due to their unique karst geology and flora (Junaid 2014; Waluyo et al. 2005). These sites overlap with primate habitat but are fewer in number and less popular than marine-oriented tourism sites. The five terrestrial national parks on Sulawesi all support tourism, though the revenue generated from tourism at these sites remains limited (Table 7.3). Bantimurung-Bulusaraung National Park generates the most income, likely due to its proximity to the capital of the Maros Regency and to Makassar, the capital of South Sulawesi province (Table 7.3). Though research on nature tourism in South Sulawesi remains limited, some evidence suggests that tourism in this region can create conflict in communities and may pose threats to biodiversity (Kadir et al. 2013; Putri 2016; Putri et al. 2020; Wakka et al. 2015). Similar concerns have been noted as tourism expanded in North Sulawesi, placing constraints on local facilities and causing environmental degradation (Hakim et al. 2012).

Table 7.3 State revenue from the utilization of the tourism potential of conservation areas in Sulawesi (Direktur Jenderal KSDAE 2016)

Conservation area management agency	Province	Revenue 2016 (IDR)
BTN Bunaken	North Sulawesi	352.023.000
BKSDA Sulawesi Utara	North Sulawesi	720.247.500
BTN Bogani Nani Wartabone	Gorontalo	30.704.000
BBTN Lore Lindu	Central Sulawesi	210.505.000
BTN Kepulauan Togean	Central Sulawesi	4.875.000
BKSDA Sulawesi Tengah	Central Sulawesi	43.490.000
(BBKSDA Sulawesi Selatan)	West Sulawesi	0
BTN Bantimurung-Bulusaraung	South Sulawesi	2.329.771.500
BTN taka Bonerate	South Sulawesi	187.694.000
BBKSDA Sulawesi Selatan	South Sulawesi	556.003.000
BTN Wakatobi	Southeast Sulawesi	71.445.000
BTN Rawa Aopa Watumohai	Southeast Sulawesi	9.045.500
BKSDA Sulawesi Tenggara	Southeast Sulawesi	7.870.000
Total revenue of all central government conservation agencies in Sulawesi		4.523.673.500

7.3.4 Primate Tourism in Sulawesi

Much like nature-based tourism more generally (Hakim et al. 2012), primate tourism predominates in Sulawesi's north province. Tangkoko Duasudara Nature Reserve (North Sulawesi) is the primary area for primate tourism in Sulawesi, largely due to the ease at which visitors are able to observe the resident primates, the Critically Endangered crested black macaque (*Macaca nigra*), and the Vulnerable spectral tarsier (*Tarsius spectrum or T. spectrumgurskyae*), which have high densities in the reserve (Arrijani 2020; Kinnaird and O'Brien 1996). Studies conducted at Tangkoko have found that primate tourism surpasses birdwatching as the primary reason for visits by foreign tourists (IUCN 2008; Kinnaird and O'Brien 1996; Sumarto and Tallei 2010). The island of Buton in Southeast Sulawesi is another site for primate tourism, specifically "research tourism" run by a UK-based conservation research organization, Operation Wallacea (Galley and Clifton 2004), whereby volunteers participate in seasonal research and conservation activities, including work on Buton macaques (*M. ochreata brunnescens*) and tarsiers (*T. spectrum*) (www.opwall.org). Aside from Tangkoko and Buton, there are few other sites where tourists can regularly encounter and easily observe Sulawesi's primates in the wild. This is largely due to a lack of tourism infrastructure (i.e., tourist facilities, tourist marketing, habituation of primates, etc.) in other protected areas that provide habitat for Sulawesi's primates. Those that do exist, such as the Karaenta area of Bantimurung-Bulusaraung National Park (see below), are not well-known as primate tourism sites, and hence, are best described as sites of "incidental tourism" (Grossberg et al. 2003; Sengupta and Radhakrishna 2020); that is, other features and activities, such as birdwatching, hiking or cultural sites, serve as the primary attraction, but once tourists are there they may have an opportunity to observe primates as well.

7.4 Incidental Primate Tourism in South Sulawesi

7.4.1 Bantimurung

The Bantimurung waterfall site is among the most well-known tourist attractions in South Sulawesi. Located within the bounds of Bantimurung-Bulusaraung National Park (TNBABUL), tourists are drawn to this outdoor recreation area to see and swim at the site's large waterfall (Rahbiah et al. 2016). In recent years, site managers have expanded the swimming areas, added a zipline above the swimming pools, and built an aerial rope bridge that offers tourists a "bird's eye" view of the recreation area (K. Morrow, personal observation). The site also hosts a museum with butterfly specimens, an outdoor butterfly park, and a large cave with notable bat roosts that tourists frequently explore. Bantimurung receives far more tourist traffic than other areas of South Sulawesi. In 2010 alone, 600,000 tourists visited the popular waterfall destination (Rahbiah et al. 2016). In this same year, only ~53,000 tourists—around 400 of which were international tourists—visited the Bone regency (Junaid 2014). Between 2009 and 2013, 2.7 million domestic tourists and 15.5 thousand international tourists visited this popular waterfall destination (Rahbiah et al. 2016).

Although the large primate statue at the site's entrance (Fig. 7.4) suggests that primates can be viewed here, employees at the recreation area report only occasional macaque sightings (Morrow 2018), and there is no established primate tourism. While portions of TNBABUL are designated as tourism forests, facilities to support tourists in this area (e.g., lodging) are limited and nearby communities may conflict with park managers regarding collaborative park management and appropriate forest use (Kadir et al. 2013; Wakka et al. 2015). However, local communities do receive some economic benefits from the tourism at Bantimurung, including by selling souvenirs at the entrance to the waterfall recreation area (Putri et al. 2020; Rahbiah et al. 2016). Butterfly specimens are often sold as souvenirs, potentially posing a risk to their conservation (Putri 2016). This site is primarily visited by local and domestic tourists, and large crowds often gather on weekends and holidays (Authors, personal observation). There is some evidence to suggest that the presence of crowds at this site negatively impacts bird communities (Putri et al. 2020).

Fig. 7.4 Authors A. Zak, K. Hanson, and E. Riley posing with the primate statue at the entrance of the Bantimurung waterfall recreation area. Photograph by IskandarKamaruddin

7.4.2 *Karaenta*

Karaenta is located in TNBABUL, a 43,750 ha park that was gazetted to protect the area's limestone (karst) ecosystem, endemic flora and fauna, and watershed systems. Karaenta was formerly a 1000 ha nature reserve that became subsumed within the national park's boundaries when the latter was established in 2004. Situated at approximately 300 m.a.s.l, Karaenta consists of primary and secondary forest amidst and upon karst tower formations that rise up to 70 m from the ground (Albani et al. 2020). Beginning in the 1980s, this area has been the primary location for ecological and behavioral research conducted on the Endangered moor macaque, *Macaca maura* (Albani et al. 2019, 2020; Germani 2016; Hanson and Riley 2017; Matsumura 1991, 1998; Morrow et al. 2019; Okamoto et al. 2000; Riley et al. 2014; Sagnotti 2013).

In the early years, researchers regularly provisioned the macaques in an effort to accelerate habituation, obtain group counts, and identify individuals (Okamoto et al. 2000; Watanabe and Matsumura 1996). Provisioning is the deliberate offering of food, typically human foods, to animals. Because macaques became well-habituated to humans at this site, particularly one group (Group B), it also became the primary location for tourists and the media to view this primate species. It is worth noting that a major road traverses through 11 km of TNBABUL, specifically through the Karaenta area, bisecting the habitat of resident fauna, including the moor macaque. Moor macaque groups have been observed crossing the road, but generally retreat back into the forest after crossing (Authors, personal observation). Accordingly, up until recently most observations of the macaques by tourists or the media were conducted inside the forest at a designated provisioning site. However, beginning in 2015, a shift occurred at this site, whereby the main habituated group (Group B) began spending more time close to the road, which in turn made them more visible to people passing in cars, at which point people began feeding them. By 2016, we estimated that group B was spending approximately 20% of the day along the road where they forage in trash pits and wait for motorists to toss them food (Morrow et al. 2019). By 2018, we observed additional groups waiting on the side of the road for provisions along the 11 km stretch through the park (E. Riley, personal observation). Therefore, opportunities for encounters with the macaques have expanded beyond the forest to include locations all along the roadside and these encounters frequently involve provisioning of anthropogenic foods (e.g., boiled corn ears, sometimes already consumed by people, chips and other snack foods, fruit such as bananas and oranges, and bread and cookies).

7.4.3 UNHAS' Hutan Pendidikan

The Hasanuddin University (UNHAS) Hutan Pendidikan (Education Forest, EF) is located in the village of Bengo and managed by the Faculty of Forestry at Hasanuddin University in Makassar. The EF serves as a teaching resource and research site for local and visiting university students and faculty. Multiple buildings provide indoor lodging for large groups and there is additional space for tent camping. Certain trees non-native to Sulawesi, such as species of pine (*Pinus merkusii*) and acacia (*Acacia* spp.), were planted in the EF during a restoration project that occurred several decades ago. Agricultural areas, including wet rice fields, mixed gardens, and cacao plantations, occur within the EF and along the eastern edge. Farmers constantly move between the village, agricultural spaces, and the forest where subsistence hunting and trapping of small game and collecting resources such as candlenuts (*Aleurites moluccana*), honey, firewood, timber, and other building materials occur

(Zak and Riley 2017). Previous research has documented at least seven groups of free-ranging moor macaques living in the EF (Agustinus. 2011). Many human-macaque interactions in Sulawesi occur predominantly at the forest-farm interface (Priston et al. 2012; Riley 2007a; b). Thus, the EF represents yet another example of a forest-farm mosaic within which local residents and researchers have confirmed the occurrence of macaque crop feeding at the forest boundary (Zak and Riley 2017; Morrow 2018). While perceptions of crop feeding behaviors are negative, farmer-macaque interactions mostly involve nonlethal deterrence methods such as the use of noisemakers and human and canine guards to chase macaques back into the forest. Retaliation killings of crop feeding macaques appear uncommon (Zak and Riley 2017; Zak 2016). There is no formally established primate tourism at this site, but it hosts occasional multiday events (e.g., the 2014 Musik Hutan, an annual music festival), UNHAS and other academic programs, and UNHAS forestry student training, including some student research on moor macaques in the EF (Agustinus. 2011). Macaque sightings are incidental and occur more rarely than in the nearby Karaenta because social groups remain relatively unhabituated and there are no paved roads that bisect the forest offering easy access for primate observation.

7.4.4 Pattanuang

Pattanuang is popular among local tourists for caving (Waluyo et al. 2005). Although Pattanuang is not a site of official primate tourism, there is a small collection of tarsiers (*Tarsius tarsier*) that are kept in outdoor enclosures in close proximity to villagers' homes and cared for by TNBABUL park rangers that park staff will occasionally show tourists (K. Morrow, personal observation; Putri 2020). These tarsiers were captured from TNBABUL forests by TNBABUL park staff and are maintained in enclosures with vegetation that allows for vertical clinging and leaping, but not cavity dwelling, during daytime hours (K. Morrow, personal observation). As of Summer 2017, TNBABUL staff had begun habituating one group of forest-dwelling tarsiers near Pattanuang by imitating tarsier vocalizations and provisioning the tarsiers with insects as they emerged from their sleeping tree in the evening (K. Morrow, personal observation). Recent research suggests that Pattanuang is of particularly high value for developing tarsier-focused tourism, but that community capacity, tourism facilities, and intentional marketing are needed to realize this potential and raise tourist interest in tarsiers (Putri 2020).

7.4.5 Bira Beach and Le'ja Hot Springs

Bira Beach in Bulukumba and Le'Ja Hot Springs in Soppeng are two additional sites where tourists may encounter macaques, though visitors are largely attracted to the destinations for marine tourism. Bira Beach is among the most popular tourist attractions in Bulukumba; in 2015, 156,770 local tourists and 3680 foreign tourists visited (Maryono et al. 2019). Government authorities have recognized the area as valuable for further developing tourism industries, but issues of waste management, improper construction of facilities, and potential overcrowding pose barriers to the feasibility and sustainability of expanding tourism (Maryono et al. 2019; Nur et al. 2019). Although visitors report seeing macaques in these areas in close proximity to humans (L. Germani, personal communication), at the time of writing this chapter, there were no research publications discussing primate tourism or human-primate encounters at these sites.

7.5 Primate Tourism in South Sulawesi Compared to Other Areas in Indonesia

The opportunities for primate tourism in South Sulawesi described above differ substantially from popular primate tourism settings in Indonesia, such as in Bali, Borneo, and Sumatra (e.g., Fuentes et al. 2007; Russon and Susilo 2014). Several factors account for these differences. First, as noted above, the sites in South Sulawesi are best understood as examples of incidental tourism (Grossberg et al. 2003; Sengupta and Radhakrishna 2020). For instance, the site at Bantimurung primarily attracts tourists for its outdoor recreation and landscape features, such that some visitors have been surprised to learn that macaques inhabit the surrounding forest (K.Hanson, unpublished data). In contrast, primate tourism sites in Bali are advertised as primate tourism destinations, have established formal tourism management and revenue-generating structures, and attract a significant number of visitors who are primarily interested in viewing primates (Fuentes 2010). Given the apparently common occurrence of incidental primate tourism in South Sulawesi, it is interesting to consider whether these types of primate encounters are prevalent outside of a formal tourism context throughout Indonesia, as appears to be the case in primate habitat countries elsewhere (Sengupta and Radhakrishna 2020).

Second, primate tourism in Indonesia is generally characterized by one of two contexts: temple settings (e.g., Bali) and forest (e.g., Kalimantan). The Sulawesi sites we have described herein fall somewhere along this spectrum. In the Karaenta area of TNBABUL, opportunistic encounters with macaque groups currently occur along the road, and hence, tourists do not need to enter the forest to observe the macaques. Likewise, tourists at the Bantimurung site can observe macaques foraging in the canopy from the more developed, waterpark-like area below. However, even macaques encountered along the road still spend the majority of the day in the forest (Morrow et al. 2019), thus providing opportunities for forest-based encounters. Though the tower karst habitat in these areas is certainly deserving of tourist appeal, navigating this landscape is challenging (see Albani et al. 2020; Hanson and Riley 2017) and has perhaps hindered the development of a forest-based model of primate tourism. Nevertheless, there are opportunities to develop forest-based primate tourism in this region, which could cater to tourists seeking more "adventurous" outdoor experiences. The small number of foreign tourist visits to Sulawesi—particularly South Sulawesi (Table 7.1)—have likely also hindered the development of primate tourism on the island. Bali receives the majority of foreign tourists, which potentially helps support established primate tourism sites. Similarly, more foreign tourists travel through North Sulawesi than other provinces, and primate tourism is more firmly established in this region.

We also suggest that the envisioned tourist experience in Sulawesi is fundamentally distinct from more popular tourist destinations in Indonesia. As discussed, Sulawesi's tourism infrastructure is not well-developed, and from a foreign perspective is more likely to attract visitors seeking an "off the beaten path" experience. Among domestic tourists, a trend toward nature-based, adventure-seeking activities makes Sulawesi an ideal destination (Butarbutar and Soemarno 2013). Taken together, these qualities should enhance the appeal of Karaenta's karst forest as a site for tourists seeking an "authentic" experience with "wild" macaques in "pristine" nature (Duffy 2002; West and Carrier 2004; Curtin 2010). In fact, developing primate tourism with this objective in mind has been expressed on several occasions by TNBABUL management, who hopes to attract domestic and foreign tourists to Karaenta to observe recently habituated moor macaques (K.Hanson, unpublished data; see Hanson and Riley 2017). What follows is a discussion of considerations as they relate to implementing a more deliberate primate tourism program in South Sulawesi.

7.6   Considerations for the Development and Management of Primate Tourism in South Sulawesi

7.6.1   Raising Awareness of Primate Biodiversity and Conservation

Conservationists often suggest that wildlife tourism expands visitors' science-based knowledge of wildlife and generates greater concern for conservation (Ardoin et al. 2015; Ballantyne et al. 2007; Powell and Ham 2008). However, existing evidence does not always support this idea, and outcomes seem to vary based on the situation and visitors' pre-existing knowledge (Hayward & Rothenberg, 2004; Hughes et al. 2011; Powell and Ham 2008). Given that domestic tourists around Karaenta and TNBABUL are often unaware that moor macaques live in the region (K.Hanson, unpublished data), it is possible that thoughtful development of primate tourism in South Sulawesi could raise awareness of local primate biodiversity and conservation. Encountering animals in contexts perceived as "natural" can lead to more positive tourist experiences and greater learning outcomes (Desmond 1999; Johnston 1998; Shettel-Neuber 1988). Thus, forest-based wildlife tourism has the potential to be a powerful means of conservation education because of encounters' high levels of perceived "naturalness" (Higham and Shelton 2011; Knight 2009, 2011).

Macaque tourism sites are not typically known for their conservation education efforts. Instead, the primary attraction for tourists at macaque sites is entertainment, while conservation and education goals are secondary or nonexistent (Knight, 2011). Establishing conservation education as part of primate tourism in South Sulawesi would therefore offer a valuable contrast to other macaque tourism sites. However, there remains extensive debate on whether and how education initiatives lead to conservation outcomes (e.g., Freund et al. 2020; Jacobson 2010; Kling and Hopkins 2015; Sherrow 2010). Implementing an education research framework— i.e., evaluating the efficacy of tourism and education initiatives before, during, and after program implementation—offers one route to establishing effective conservation programming (Padua et al. 2002; Sherrow 2010; Yu et al. 2011). In addition, the success of programs may greatly depend on how information is communicated or messaged. Historically, interpretative signage has been used to convey information in science-related settings, but venturing beyond simply passing on information is critical (Orams 1994).

The relatively new opportunities to encounter macaques along the road in this region (Morrow et al. 2019) underscore the potential value of formally establishing primate tourism and conservation education. Motorists passing through Karaenta have often encountered moor macaques (*Macaca maura*) in crop-foraging contexts and, as a result, view cultivated foods as typical macaque dietary resources. Furthermore, the macaques' physiological and behavioral similarity to humans and their dietary flexibility leads people to perceive anthropogenic processed foods as suitable items to provision the monkeys (Morrow 2018). Formal primate tourism with established educational components may help encourage more responsible human-macaque interactions at this site and reduce the instances of provisioning the macaques with processed foods. Education components could address conservation concerns related to provisioning and the role that macaques play in local ecosystems (Tsuji and Su 2018).

Establishing primate tourism in South Sulawesi would also introduce employment opportunities and formal management of the human-primate interface, which may enhance local community members' awareness and support of conservation efforts. Research at several wildlife tourism sites suggests that involving community members—for instance, as guides—can positively impact community conservation knowledge and attitudes (Keane et al. 2011; Waylen et al. 2009), increase success of conservation initiatives (Waylen 2010), provide opportunities for local communities to share their own culture and knowledge systems (Zeppel and Muloin 2008), and benefit the health and wellbeing of participants (Moore et al. 2006). Importantly, focusing only on the economic benefits of primate tourism may not lead people to change their conservation attitudes and behaviors (Nilsson et al. 2016; Stem et al. 2005). Rather than relying solely on these extrinsic motivators, designing tourism and conservation programs that focus on intrinsic motivators (e.g., caring for the environment) is more likely to result in sustainable conservation outcomes that benefit primate habitats (Nilsson et al. 2016). Knowledge, support, and success of conservation are especially likely to result when local community members are involved as significant stakeholders with autonomous management and decision-making authority (Stronza and Pêgas 2008; Waylen 2010). In South Sulawesi, it may therefore be beneficial for protected area staff, community members, and researchers to co-develop tourism programs that provide economic benefits, emphasize the intrinsic values of conservation, and are managed and maintained by community stakeholders.

7.6.2 Ethical Dimensions of Primate Tourism

In considering developing primate tourism in South Sulawesi, several ethical dimensions arise. In what follows, we briefly discuss the four most prominent ethical considerations: the biological, ecological, and behavioral impacts of tourism on primates, the ethics of habituating wild primates, the potential for exacerbating macaque crop foraging, and the impacts on local livelihoods.

7.6.2.1 Biological, Ecological, and Behavioral Impacts of Tourism

Despite wildlife tourism's potential contribution to the conservation of biodiversity, including primate conservation, there is a growing concern regarding the impact of tourism on primate ecology, behavior, and health (Fuentes and Gamerl 2005; Ilham et al. 2018; Maréchal et al. 2016; Russon and Wallis 2014) as well as on the ecosystem as a whole (Larson et al. 2016; Shannon et al. 2017). Although nature-focused tourism initiatives are almost certainly less damaging than more invasive land-use practices (e.g., extractive industries), some of the main concerns regarding the ecological effects of tourism include habitat modification and human habitat use, which can result in animal behavioral shifts and physiological disturbances (Buckley 2004). Habitat modification can be defined as any alteration to the environment by humans (i.e., trails, barriers, sounds, smells, ground cover or water source removal). Effects from habitat disturbances may ecosystem dynamics and interspecies interactions in complex ways, particularly for species whose resource dependence varies with age or seasonal availability (e.g., Morgan et al. 2018). Further research is needed to better understand how the ecological impacts of tourism affect primates, particularly as many primate species perform vital ecosystem services (Trolliet et al. 2016). In South Sulawesi, it may be especially important to assess the effects of tourism on macaque feeding ecology; as one of the largest mammals and frugivores in the region, they likely play an important role in the ecosystem as a whole (Tsuji and Su 2018; Whitten et al. 2002). Monitoring the effects of tourism on the more ecologically and behaviorally specialized tarsiers will also be important, although research from other tourism sites suggests that they may be resilient to tourism activities (Paulus 2009). Any development of primate tourism in South Sulawesi should also consider the potential ecological impacts on other organisms. Of particular concern are Bantimurung-Bulusaraung National Park's notable butterfly species, which may already be negatively impacted by tourism (Putri 2016) and the endemic cuscus (*Ailurops ursinus*), which may be hunted for consumption by local communities (Salas et al., 2019).

In contrast to indirect ecological impacts, direct behavioral consequences of primate tourism are more widely studied. For example, the presence of tourists and the behaviors they exhibit when around wild primates have been shown to increase stress among primates, as measured by rates of anxiety-related behavior, such as self-scratching, and physiological measures of stress, such as fecal glucocorticoid concentrations (e.g., Barbary macaques; Maréchal et al. 2011; Black howler monkeys; Behie et al. 2010). One of the most common ways humans and primates interact at tourist sites is through provisioning (Sengupta and Radhakrishna 2020). Provisioning affects primate feeding ecology, such as reducing dietary diversity (Sengupta and Radhakrishna 2018), as well as ranging behavior. For example, a number of studies have shown that provisioned primates show smaller home ranges and shorter daily travel distances, a pattern which likely reflects how the high abundance and clumped distribution of provisioned foods at these sites reduce travel costs for the primates (Hansen et al. 2020; Sengupta et al. 2015). Provisioning has also been shown to influence primate social behavior. For example, provisioning can result in increased intragroup aggression (Hsu et al. 2009; Ram et al. 2003), reduced time spent grooming (Kaburu et al. 2019), reduced social cohesion (Morrow et al. 2019), and changes in group size (Riley et al. 2016) and social structure (Sinha et al. 2005).

Provisioning primates at tourism sites also poses a serious risk of pathogen transmission (Carne et al. 2017; Sapolsky 2014). Bidirectional pathogen transmission between humans and other primates is a significant conservation concern and an important ethical consideration for developing and managing primate tourism (Fuentes 2006; Jones-Engel et al. 2005; Muehlenbein 2017). While all wildlife tourism sites must grapple with the potential for pathogen transmission, the risk of pathogen transfer in the context of primate tourism is heightened due to our close phylogenetic relationships and similar biology (Olival et al. 2017; Wallis and Lee 1999). Furthermore, primate populations are especially susceptible to disease due to their long, slow life histories, which hinder rapid recovery from population declines (Dunbar 1987; MacArthur and Wilson 1967; Purvis et al. 2000; Ross 1992). Suggested steps to mitigate pathogen transmission at primate tourism sites include limiting tourist attendance, complying with rules and regulations for maintaining safe proximity to primates, implementing health education programs, vaccinating both humans and nonhuman primates, and following appropriate behavioral hygiene guidelines, including wearing face masks, maintaining adequate distances from primates, and prohibiting symptomatic visitors and staff from participating (Homsy 1999; Russon and Wallis 2014; Ryan and Walsh 2011).

Mitigating the risk of human-macaque pathogen transfer in South Sulawesi may be particularly challenging. As with other macaque tourism sites (e.g., Brotcorne et al. 2017; Carne et al. 2017; Fuentes and Gamerl 2005; Hsu et al. 2009; McCarthy et al. 2009), sites of incidental macaque tourism in South Sulawesi involve provisioning and close proximity to humans (Morrow et al. 2019). However, these interactions are generally unmanaged in South Sulawesi. Along the Bira cape and in the

Karaenta area of TNBABUL in South Sulawesi, passing motorists often provision moor macaques (L. Germani, personal communication; Morrow et al. 2019). Such practices are especially risky for tourists if they involve scratching and biting by the macaques, as macaques are known to carry the Herpes B virus, which can be fatal in humans (Engel et al. 2002). Interview data suggest that people feel it is acceptable to feed moor macaques in South Sulawesi for a number of reasons: it is a common occurrence at well-known long-tailed macaque tourism sites in Bali; the national park and researchers have historically provisioned the monkeys; and, the macaques appear to be "hungry" and actively looking for human food (Morrow 2018). These existing perspectives may make managing the risks of provisioning in this region difficult. Indeed, evidence from other tourism sites suggests that people often do not follow established regulations and will still seek direct contact with primates even when they understand the potential for pathogen transmission (Nakamura & Nishida, 2009; Muehlenbein et al. 2010).

In addition to pathogen transfer risks, other negative health outcomes associated with provisioning includexsdz food poisoning (Maréchal et al. 2016), fatal ingestion of inappropriate foods or objects (Rodriguez-Lopez and Mignucci-Giannoni 1999), and increased rates of obesity, which can lead to reduced fertility and other nutrition-related health concerns (Sapolsky 2014). Given the suite of negative impacts outlined above, any efforts to develop new sites of primate tourism in South Sulawesi should avoid, or more preferably, prohibit provisioning. While it may be difficult or impractical to eliminate primate provisioning in South Sulawesi at sites where it is already occurring, we recommend continued outreach focused on augmenting people's knowledge and understanding of the negative consequences of provisioning and how just because it occurs elsewhere in Indonesia does not mean it is a good model for Sulawesi. Potential primate tourism sites should also prioritize reducing negative ecological impacts and work with collaborators to conduct continuous research on ecosystem health.

7.6.2.2 The Ethics of Habituating Wild Primates

The ethical imperative to "do no harm" (Riley and Bezanson 2018) is complicated with respect to habituation, because "harm" can also occur in less conspicuous ways. Knight (2009) identifies two methods of rendering wildlife "viewable" to tourists: habituation and attraction (i.e., via provisioning, as discussed above). Indeed, habituation is distinct from tolerance attained through provisioning; the latter, some have argued, is best understood as associative learning (Bejder et al. 2009; Higham and Shelton 2011). While both habituation and attraction represent a heightened tolerance of human observers, it is a long-held assumption that habituated primates perceive humans as a neutral presence and no longer respond to them, or that human presence is not disruptive (Allan et al. 2020; Fedigan 2010; Higham and Shelton 2011; Tutin and Fernandez 1991; Williamson and Feistner 2011). Recent work in this area, however, has challenged this accepted premise (Alcayna-Stevens 2016; Allan et al. 2020; Ampumuza & Driesson, 2020; Hanson and Riley

2017). Not only do habituated primates continue to respond past the point of what observers might consider "habituated," but they do so in ways that may go undetected or are only revealed through long-term monitoring and analysis (Bejder et al. 2009; Hanson and Riley 2017; Higham and Shelton 2011; McDougall 2012). Here, we adopt the view of habituation as a dynamic and context-dependent spectrum of heightened observer tolerance (see Hanson and Riley 2017).

Deploying this nuanced understanding of habituation has important implications for primate tourism. Though it is convenient to presume that a habituated primate group is "immune" to day-to-day observer influence, multiple daily follows with tourist groups over time have the potential to induce a chronic stress response that can ultimately impact the groups' wellbeing (Chen et al., 2020; Hanson 2017; Shutt et al. 2014). Other research demonstrates that persistent following of presumed habituated groups is associated with increased locomotion and decreased resting periods (Rassmusen 1998; Hanson 2017). Moreover, evidence suggesting that the habituation process results in differing tolerance levels across individuals (Allan et al. 2020; Ampumuza & Driesson, 2020; Bertolani and Boesch 2008; Narat et al. 2015) points to the possibility that tourist impact is not uniformly distributed across individuals and across social groups (Allan et al. 2020; Morrow et al. 2019; Westin 2017). For example, individuals with higher observer tolerance could potentially use humans as social tools for accessing and retaining food resources or avoiding aggression (Allan et al. 2020, p. 10; Hanson and Riley 2017). Precautions to mitigate these impacts, such as "no research" policies on groups habituated for tourism and limiting tourist group size and visits to one hour per day, are already incorporated into best practice guidelines for great ape tourism (Williamson and Macfie 2014). Other management strategies that approach habituation (and hence, its consequences) as a flexible spectrum may seek to structure the nature of tourist-primate interactions at the level of the individual animal (Higham and Shelton 2011, p. 1296; Ampumuza & Driesson, 2020).

For the ethically driven primatologist, upholding the principle of "do no harm" is a given. A recent survey conducted by Green and Gabriel (2020) confirms that primatologists feel a strong duty to mitigate research and other human-caused harms to their habituated study subjects, but we must also be careful that observer tolerance and the assumptions it entails do not obscure harm—subtle or otherwise. An important question that emerges from Green and Gabriel's (2020) analysis is whether habituation is necessary for primate tourism. For those tourists who seek nature-based excursions in South Sulawesi, perhaps hiking through an appealing forest for a glimpse of unhabituated macaques is enough to satisfy their appetite for adventure.

7.6.2.3 Macaque Crop Foraging

Crop feeding is a widespread problem across primate ranges and Sulawesi is no different. All seven macaque species are believed to engage in the behavior (Riley 2010). Farmers working within and around the UNHAS Education Forest (EF, see above) in South Sulawesi report that crop feeding leads to reduced harvests which result in smaller incomes (Zak 2016). Additionally, the most effective deterrence method for protecting a garden, human guarding (Cai et al. 2008; Hill 2000; Nijman and Nekaris 2010; Zak 2016), is a time-consuming job that prevents farmers and their families from engaging in other tasks beneficial to their livelihood (e.g., finding honey to sell), and may result in health risks such as lack of sleep if guarding at night, exposure to dangerous animals and diseases (e.g., dengue fever, malaria), and children missing school (Osborn and Hill 2005). Deterrence method efficacy is also affected by factors outside of farmers' control. For example, the practice of provisioning primates may exacerbate the issue of crop feeding by increasing the likelihood of crop damage and influencing primate reactions to various deterrence methods (Madden 2006).

The decision to habituate macaques that live in forest-farm matrix habitat comes with practical and ethical concerns. First, it may be more difficult to habituate a group of primates that have had repeated negative encounters with farmers. Additionally, habituating a group that will potentially crop feed may lead to decreased fear of human guards (Fuentes and Hockings 2010) and increased conflict with humans (McLennan and Hill 2010) in agricultural spaces. From a conservation perspective, habituating Sulawesi macaques for tourism may also lead to increased retaliatory killings or harm to an endangered species, as individuals that are more accustomed to being in proximity to humans are easier to capture and punish (Zak 2016). While these concerns entail a view of habituation characterized by the loss of fear of humans, we have argued above that habituation is deeply situational, and hence, it may be unrealistic to presume that macaques will extend this loss of fear to all humans and in all contexts. Nevertheless, we recommend seriously considering the risks in habituating social groups that range in close proximity to agricultural spaces, because the factors influencing crop feeding are interrelated in complex ways (Hill 2018). This recommendation is not to undermine macaque capabilities to read various contexts and respond accordingly, but rather is suggested out of caution and respect for both human and nonhuman primate communities. Reducing future conflict can be achieved in part by preventing noncrop feeding groups from learning to do so. Furthermore, although we fully support collaborations between Western and Indonesian researchers, working with universities in the region should not be conflated with working with local people (Lowe 2004). In fact, local communities and forest managers may have drastically different ideas about how the forest has been and should be used and the status of wildlife within. For example, although UNHAS manages the Education Forest in a rural village, their faculty and staff do not necessarily represent the perspectives of Bengo farmers and residents. While buffer crops and other examples of intentional provisioning have been suggested to mitigate crop loss (Hockings and McLennan 2012; Parker and Osborn 2006; Riley 2007b; Zak, personal communication), provisioning macaques within the EF for research or tourism could potentially complicate relationships between university staff, researchers, and the local community, particularly if farmers are not involved in the decision making and have concern that activities might increase crop feeding behavior as habituated primates lose their fear of humans. A positive relationship and effective communication between forest managers and local communities is critical to ensure the protection of existing macaque populations while allowing people to continue to farm, engage in responsible tourism, and use forest resources in ethical ways.

7.6.2.4 Local Livelihoods

Economic impacts associated with primate tourism are well-documented (see Hvenegaard 2014 for overview), with outcomes varying considerably from site to site (Eshun and Tonto 2014; Wright et al. 2014). Primate tourism's economic contribution largely depends on the degree to which initial tourist spending remains in the region (Hvenegaard 2014). Thus, if revenue outflow is high, it is unlikely local communities and protected areas will benefit from primate tourism. This was the case in Tangkoko Duasudara Nature Reserve, Northeast Sulawesi where Kinnaird and O'Brien (1996) reported that the local community did not profit from tourist visits and only 2% of tourist revenues remained in the reserve. In South Sulawesi, a more recent study examining the role of Bantimurung tourism in local livelihoods indicates potential economic and social benefits, including income diversification, increased monthly income, and enhanced opportunities for female employment (Rahbiah et al. 2016). The majority of these benefits are derived from centralized activities at the entrance of the park, such as selling souvenirs or snacks (Ibid.). However, it is debatable whether primate tourism in Bantimurung would draw more tourists than usual and thus augment local economies and livelihoods in a significant way.

Though many have argued that economic benefits from primate tourism promote local support for conservation objectives (Russon and Wallis 2014), others have challenged the assumed connection between economic incentive and conservation success (Stronza 2007; Fletcher 2009). Articulating with this critique is the idea that environmentally and socially responsible tourism (i.e., ecotourism, see Stronza et al. 2019)paradoxically functions as a "capitalist fix" to redress environmental and social ills caused by capitalist development (Fletcher and Neves 2012; Büscher et al. 2012). Thus, the development of primate tourism as a market-based conservation strategy has the potential to exacerbate existing social inequities and actually hinder long-term conservation efforts (West 2006; Duffy 2013). Further, without empirical or situated evidence, we cannot assume increased local income from primate tourism will lead to increased conservation (Fletcher 2009). For instance, Stronza (2007) describes a dynamic in which increased income from ecotourism enables and accelerates resource extraction due to local residents' newfound ability to purchase labor and technology. Primate tourism's impact on local livelihoods and conservation must therefore be assessed in light of other extractive and arguably more harmful industries in South Sulawesi (e.g., large-scale agriculture, nickel mining; Supriatna et al. 2020). Finally, conflict over which entities profit from tourist activity and revenue instability related to tourist seasonality, economic trends, and recently, global health crises (Dinarto et al. 2020) can complicate and undermine positive contributions of primate tourism to local livelihoods. Ensuring that economic benefits from tourism activities are sustainable, equitable, and transparent is not straightforward and would necessitate open communication between local community members, park rangers and managers, and researchers.

7.6.3 Who Are the Tourists?

Tourist demographics can also play a key role in the development of primate tourism. In South Sulawesi, there is limited tourist-focused infrastructure surrounding protected areas where primate viewing occurs (see above). Evidence from popular primate tourism sites in North Sulawesi demonstrates that primate tourism is especially popular with international tourists (Kinnaird and O'Brien 1996), pointing to South Sulawesi's attractive potential. Furthermore, at a macaque tourism site in Padangtegal, Bali, non-Asian tourists comprised 50% of total visitors (Fuentes et al. 2007). Even though Bali is known as the international tourist hot spot of Indonesia, the presence of Western tourists at primate-focused localities throughout Indonesia indicate that international interest in primate tourism may translate, to a degree, to South Sulawesi.

Because the majority of tourists who arrive in Sulawesi are not international, it is likely that local and domestic tourists will play an important role in shaping primate tourism in the region. As such, it is important to understand domestic tourists' motivations for participating in nature-focused tourism. Whereas Western tourists tend to view nature-based tourism as a way to quietly appreciate and reflect on nature, Indonesian "nature-loving" dates to Suharto-era periods of political suppression and is steeped in ideals of nationalism (Collins 2007; Tsing 2005). Accordingly, Indonesian nature-loving prioritizes "conquering" nature by taking group adventures to isolated, dramatic vistas, rather than traveling alone or in small groups to experience and learn about nature (Tsing 2005). At less isolated tourism sites, domestic tourists primarily visit for recreation and social engagement, rather than to learn about the conservation status of an ecosystem (Cochrane 2006). Therefore, potential primate tourism sites in South Sulawesi may be expected to cater to large groups of tourists who are more interested in brief encounters with primates rather than prolonged encounters that emphasize educational programming. Finding ways to balance conservation education goals while meeting domestic tourists' desired experiences will thus require creative planning.

7.6.4 What Is the Role of the Researchers?

As anthropologically trained and ethically engaged practitioners, the degree to which the academic researcher is involved in knowledge sharing and co-development of nature-based tourism is of particular importance to us. Several examples of successful primate tourism initiatives highlight collaboration between primate researchers and local entities in designing, managing, and monitoring these programs (see Wright et al. 2014; Williamson and Fawcett 2008), yet the researcher's role in guiding and informing wildlife and nature-based tourism remains an overlooked and contentious issue (Higuchi and Yamanaka 2017; Rodger et al. 2010). Given that nature-based tourism involves the interface between society and natural resources, it is not a revelation that the development of sustainable, equitable, and responsible programs necessitates knowledge of biological, ecological, and social realms, and ideally, the synergistic relations between all three. As we have argued throughout this chapter, effective primate tourism demands productive partnership and trust among local community members, protected area managers and staff, conservation practitioners, primatologists, and other researchers. Here, we seek to encourage engagement by highlighting the valuable insights researchers contribute to thoughtful design and implementation of primate tourism. We also propose that the research informing and sustaining such programs will undoubtedly benefit from more inclusive, integrative, and transdisciplinary approaches.

A key aim for responsible primate tourism is to ensure primate wellbeing by minimizing tourism disturbance on primates and their habitats (Russon and Wallis 2014). In order to achieve this, however, there must be a foundational understanding of site-specific patterns from which to identify potential impacts, monitor and address emergent ramifications, and develop appropriate management strategies (Rodger and Calver 2005). Researchers can play an instrumental role in this regard—especially those who seek to make their research goals and questions relevant to the local communities that sustain their fieldwork. Experiential knowledge gained through fieldwork, such as daily activity rhythms and ranging patterns, can also benefit tourism design by facilitating observation conditions (Williamson and Fawcett 2008), and thereby reducing potential sources of stress for the primates while increasing tourist satisfaction (Setchell et al. 2017). In some instances, research findings and subsequent media coverage have been used as tools to promote public awareness and attract tourists (Kurita 2014; Wright et al. 2014). In Karaenta, for example, researchers have used local media interest as an opportunity to encourage conscientious encounters with the macaques and responsible human behavior (e.g., E. Riley consulted on a Mongabay Indonesia article about the risks of pathogen transmission for moor macaques; Rusdianto 2020).

Reasons for researchers' reluctance to engage in knowledge sharing and co-creation of wildlife tourism initiatives are multifaceted and never straightforward, but scholarship in recent years increasingly underscores the widespread inability to reach cooperatively and productively across disciplinary divides (Chua et al. 2020; Setchell et al. 2017; Rodger et al. 2010). At the same time, the value of integrative methods and transdisciplinary approaches for illuminating the spaces obscured by the perennial epistemological abyss cannot be overstated (Setchell et al. 2017; Riley 2013, 2019, Riley and Bezanson 2018; Fuentes et al. 2017; Remis and Jost Robinson 2020). Primate tourism as a long-term conservation strategy, an avenue for social justice and economic empowerment, and an effective education tool cannot be fully realized without inclusive, collaborative, and progressive work. Since many primate researchers are (1) not local to the habitat country and (2) not trained in social sciences, it is especially critical to reach, speak, and learn across the divide while also seeking local collaboration, so that we are all better equipped to apprehend primate tourism's benefits and risks as well as address the unique challenges it poses.

7.7 Emerging Concerns

7.7.1 Social Media Usage and Wildlife Tourism

Social media can play an important role in motivating tourists to visit nature-based tourism sites (Divinagracia et al. 2012). Although tourists frequently share their encounters with nonhuman primates on social media (e.g., Otsuka and Yamakoshi 2020), the impact of social media on wildlife tourism and human-wildlife encounters is poorly understood. Evidence indicates that emotion plays a key role in wildlife encounters (Ballantyne et al. 2007; Kellert et al. 1996; McIntosh and Wright 2017) and that many people actively seek out experiences with "wild" animals (Fuentes et al. 2007; Griggio 2015; Jones 2011), especially if it involves viewing species that are considered "charismatic megafauna" (Reynolds and Braithwaite 2001). Images and "selfies" documenting and commemorating these encounters are a socially significant component of human-wildlife interactions (Desmond 1999; Griggio 2015; Kurniawan et al. 2017) and form complex connections between tourists and places or experiences (Pearce and Moscardo 2015). As social media platforms serve as venues for sharing personal photographs and videos, they are likely relevant to understanding how people interact and seek encounters with primates in Indonesia (cf. Hausmann et al. 2018; Otsuka and Yamakoshi 2020; Tenkanen et al. 2017). The influence of social media is particularly relevant in Indonesia, which has one of the largest user bases of social media platforms worldwide (Kemp 2020) and where it is common for people to take selfies in problematic, dangerous contexts (e.g., vehicle collision incidents) (Kurniawan et al. 2017).

Although there are limited data on the role of social media in shaping wildlife tourism, there is ample evidence that media plays an important role in public perception of wildlife and of conservation. For instance, videos and images of individual animals can encourage people to want primates as pets or to want to touch animals perceived as "cute" (Chua 2018; Nekaris et al. 2013). Similarly, videos showing people in proximity to primates receive more views and responses online (Otsuka and Yamakoshi 2020), and images of primates in anthropogenic contexts can lead people to think the species represented are not endangered (Ross et al. 2008; Ross et al. 2011). There may, however, be benefits to documenting wildlife on social media. Articles or posts with images of animals may be more likely to be shared across social media platforms (Papworth et al. 2015), which could help spread information on a given species or facilitate dissemination of conservation information. Given that motorists who encounter macaques along roads in South Sulawesi report being motivated to share images of the monkeys on social media platforms (Morrow 2018), it is possible that responsible photography at South Sulawesi primate tourism sites could help raise awareness of local wildlife and conservation issues. Protected areas in South Sulawesi could play an important role in demonstrating responsible wildlife tourism photography; social media platforms maintained by protected area staff (e.g., TNBABUL Instagram account) offer an existing foundation on which to promote these ideas. However, precautions should be taken to minimize the negative conservation and perception-related consequences of social media use and to encourage responsible photography of wildlife. This can be achieved in part by modeling and disseminating recent best practice guidelines for responsible images of nonhuman primates (Waters et al. 2021), which has been translated into several languages, including Bahasa Indonesia.

7.7.2   Impacts of COVID-19

The COVID-19 pandemic has disrupted many industries, including primate tourism (Lappan et al. 2020). Stay-at-home orders, travel restrictions, and the closures of protected areas and other tourist sites have meant fewer tourists, and hence, fewer human-macaque encounters. Accordingly, rates of provisioning (Lappan et al. 2020), the likelihood of human-directed aggression (e.g., Beisner et al. 2015; Hsu et al. 2009), and the risk of zoonotic exchange (Balasubramaniam et al. 2020) potentially resulting from these encounters have also been reduced. Although these changes can be considered more positive outcomes of the COVID-19 pandemic, it is likely that they will only be temporary, and that the intensity of human-primate interactions will once again increase as travel restrictions loosen and if concerns about the risk of zoonosis decrease. On the other hand, it is also possible that heightened awareness of the risk of zoonosis due to the COVID-19 pandemic has made communities more receptive to messaging about the risk provisioning and other encounters with primates pose for human-primate pathogen transmission (Lappan et al. 2020). Finally, the COVID-19 pandemic highlights the unpredictable nature of the tourism industry and thus also the instability of economic benefits it confers (Dinarto et al. 2020). Though macaques in South Sulawesi potentially stand to benefit from the consequences of reduced tourism, the same cannot be said of the local human communities who may rely on tourist revenue as a source of income.

7.8   Conclusion: Expanding Tourism in Sulawesi

Indonesia is increasingly looking to tourism for economic development opportunities that benefit communities while protecting local culture and ecology (Junaid 2014; Kodir et al. 2020; Prodjo 2017). Tourism-based economic development may also play an important role in reducing Indonesia's reliance on other foreign exchange industries, including the top three industries of coal, gas, and oil palm

(Kodir et al. 2020). In Sulawesi specifically, promoting tourism that engenders support for protecting forested habitats may provide incentives to stem growing deforestation driven by corn, coffee, cocoa, and oil palm agriculture (Supriatna et al. 2020). Such extraction-based industries cause significant environmental damage and often benefit governments and large corporations rather than local communities (Santika et al. 2019; Welker 2014). Tourism ventures can also provide people with additional income sources and livelihood strategies when implemented equitably and sustainably (e.g., individuals who sell souvenirs at the entrance of the Bantimurung waterfall site, Rahbiah et al. 2016). Given that tourism in Sulawesi is underdeveloped compared to other areas of Indonesia (Badan Pusat Statistik 2020; Junaid 2014), there remain numerous opportunities to sustainably showcase Indonesia's biological and cultural diversity. Moreover, Sulawesi's unique ecology offers valuable nature-based tourism opportunities that differ from those found in other areas of Indonesia, including encountering endemic macaque and tarsier species in forested environments.

In this chapter, we explored the potential for primate tourism in South Sulawesi and the considerations that would be needed in developing and managing tourism initiatives focused on the islands' macaque and tarsier species. The lack of established primate tourism in this region--and in Sulawesi more generally--can likely be attributed to the historical emphasis on cultural tourism on the island (e.g., Adams 1997) and to the greater popularity of other islands (e.g., Bali) among foreign tourists (Badan Pusat Statistik 2020). However, a significant extent of land in South Sulawesi is designated as nature tourism parks (Forclime 2017), and there are a number of tourist sites throughout the province with existing incidental primate tourism where people encounter primates but there is no formal management of these encounters.

Potential tourism and conservation management programs may want to consider selecting the Sulawesi primates as flagship, umbrella, or focal species, which could help to promote the protection of the surrounding ecosystem without targeting specific species (McGowan et al. 2020; Roberge and Angelstam 2004; Wilcove 1993). South Sulawesi, in particular, is notable for its limestone karst habitats and caves (Junaid 2014; Rahbiah et al. 2016; Waluyo et al. 2005); outside of protected areas these habitats are often threatened by cement mining industries (Clements et al. 2006). Further tourism development in karst habitats may diminish this conservation threat while providing alternative income sources for local communities. The demonstrated popularity of nature-focused tourism in the Maros karst area (Junaid 2014; Rahbiah et al. 2016) suggests that such an approach may be a viable long-term conservation strategy. Similarly, there are numerous dramatic waterfall sites throughout South Sulawesi that could serve as destinations for domestic and foreign tourists alike. The popularity of better-known waterfall sites, such as the Bantimurung waterfall recreation area (Junaid 2014; Rahbiah et al. 2016), indicates that such destinations may be popular among tourists given appropriate management and facilities development. South Sulawesi's karst and waterfall sites often overlap with primate habitat. The development of nature tourism more generally in South Sulawesi could thus create additional opportunities for primate tourism in the region, which would highlight the endemic macaque and tarsier species found on the island. Likewise, expansion of tourism industries throughout Sulawesi creates opportunities for cultural tourism to showcase the island's diverse ethnic groups and cultural practices.

Expanding tourism in South Sulawesi may serve to reduce the pressure on other popular tourist destinations in Indonesia, including Bali, Jakarta, Batam, West Java, and Medan (Junaid 2014) while promoting a more equitable distribution of the economic benefits of Indonesia's tourism industry. However, tourism facilities and associated logistical resources (e.g., hotels and homestays, established tourism transportation) are currently lacking even at existing tourist sites in South Sulawesi and would need to be developed before tourism—particularly international tourism—could feasibly expand (Kadir et al. 2013; Putri 2020). Given that conflict between managers and local communities already exists at some established tourism sites in South Sulawesi (e.g., Bantimurung, Wakka et al. 2015), careful collaboration among national parks, forestry officials, communities, and researchers would be necessary to ensure transparency, sustainability, and equitable benefits sharing. Issues of sustainability are particularly relevant to potential primate tourism in the region as many of Sulawesi's macaque and tarsier species are endemic and threatened with extinction (Groves and Shekelle 2010; Merker et al. 2010; Shekelle et al. 2017; Riley 2010). Existing opportunistic primate encounters in the region already face sustainability challenges, primarily due to the issue of unmanaged provisioning of moor macaques encountered along roads in the Karaenta area of Bantimurung-Bulusaurang National Park (Morrow et al. 2019).

We suggest that with cautious, intentional development, both the people and the nonhuman primates of South Sulawesi could benefit from primate tourism. Capitalizing on the opportunities to encounter macaques and tarsiers in forested environments would offer alternatives to popular urban-based primate tourism in Bali (e.g., Fuentes et al. 2007) and could support branding Sulawesi as a tourism destination focused on distanced viewing of primates in forested habitats. Focusing specifically on tourism approaches that support primate well-being—such as avoiding provisioning primates and not habituating primates that already forage in agricultural areas—can further encourage ethical wildlife and nature-based tourism practices throughout Indonesia. Close collaboration with communities will be necessary to ensure that local livelihoods are not negatively impacted and that local people have sufficient opportunities to benefit economically from tourism, even if it occurs within the bounds of government-run protected areas. To ensure effective and culturally relevant design, implementation, and marketing of primate tourism, it will also be important to collaborate with local stakeholders or researchers who understand the perspectives and goals of Indonesian tourists, who often engage with nature differently than tourists from the global North (Cochrane 2006; Tsing 2005).

Along with these recommendations, more research is necessary to fully understand the balance of benefits and risks posed by developing primate tourism in South Sulawesi. Herein, we review a number of important facets that should be considered, including assessing the biological, ecological, and behavioral effects of tourism on primates, the ethics of habituating primates, the potential to exacerbate crop feeding by macaques, the implications for local livelihoods, and equitable management of tourism practices. Collaboration among managers, researchers, and community members will be important to develop sustainable primate tourism in this region. Emerging efforts to implement these programs should focus on meeting the desires of domestic tourists—the main visitors to Sulawesi—while promoting practices that prioritize primate wellbeing.

While the analysis we present here focuses specifically on the current state of human-nonhuman primate coexistence in South Sulawesi, Indonesia, incidental primate tourism is common throughout primate ranges (e.g., Belize; Grossberg et al. 2003 and India; Sengupta and Radhakrishna 2020). The considerations we raise in this chapter are, therefore, applicable to other regions where formal primate tourism remains underdeveloped. In particular, we suggest that researchers and practitioners working to develop primate tourism in other regions carefully consider existing dimensions of human-nonhuman primate conflict, tourism audiences, and local engagement from diverse stakeholders. Primate tourism has the potential to support the conservation of threatened primate species while also advancing economic development in primate habitat countries; however, to be effective and sustainable, the wellbeing of nonhuman primates and local communities must be prioritized across all stages, from design to implementation.

References

Act of the Republic of Indonesia No. 5 of 1990 (1990) Concerning conservation of living resources and their ecosystems. Ministry of Forestry

Adams KM (1997) Ethnic tourism and the renegotiation of tradition in Tana Toraja (Sulawesi, Indonesia). Ethnology 36(4):309–320. https://doi.org/10.2307/3774040

Agustinus (2011) Daerah jelajah dan potensi jenis tumbuhan pakan M. maura pada kelompok 1 di Hutan Pendidikan (Unpublished bachelor's thesis). Universitas Hasanuddin

Albani A, De Liberato C, Wahid I, Berrilli F, Riley EP, Cardeti G, Ngakan PO, Carosi M (2019) Preliminary assessment of gastrointestinal parasites in two wild groups of endangered moor macaques (Macaca maura) from Sulawesi. Int J Primatol 40(6):671–686. https://doi.org/10.1007/s10764-019-00114-w

Albani A, Cutini M, Germani L, Riley EP, Ngakan PO, Carosi M (2020) Activity budget, home range, and habitat use of moor macaques (Macaca maura) in the karst forest of South Sulawesi, Indonesia. Primates. https://doi.org/10.1007/s10329-020-00811-8

Alcayna-Stevens L (2016) Habituating field scientists. Soc Stud Sci 46(6):833–853

Allan ATL, Bailey AL, Hill RA (2020) Habituation is not neutral or equal: individual differences in tolerance suggest an overlooked personality trait. Sci Adv 6(28):eaaz0870. https://doi.org/10.1126/sciadv.aaz0870

Ampumuza, C., & Driessen, C. (2020). Gorilla habituation and the role of animal agency in conservation and tourism development at Bwindi, South Western Uganda. *Environment and Planning E: Nature and Space*, 251484862096650. https://doi.org/10.1177/2514848620966502

Ardoin NM, Wheaton M, Bowers AW, Hunt CA, Durham WH (2015) Nature-based tourism's impact on environmental knowledge, attitudes, and behavior: a review and analysis of the literature and potential future research. J Sustain Tour 23(6):838–858. https://doi.org/10.1080/09669582.2015.1024258

Arrijani RM (2020) Vegetation analysis and population of tarsier (*Tarsius spectrumgurskyae*) at Batuputih nature Tourism Park, North Sulawesi, Indonesia. Biodivers J Biol Diver 21(2):Article 2. https://doi.org/10.13057/biodiv/d210214

Aubert M, Brumm A, Ramli M, Sutikna T, Saptomo EW, Hakim B, Morwood M, van den Bergh GD, Kinsley L, Dosseto A (2014) Pleistocene cave art from Sulawesi, Indonesia. Nature 514(7521):223–227

Babcock TG (1982) Notes on ethnic factors related to development in Sulawesi, Indonesia. Asian J Soc Sci 10(1):116–123. https://doi.org/10.1163/156853182X00083

Badan Pusat Statistik (2014) Population projection by province, 2010–2035 (Thousand). https://www.bps.go.id/statictable/2014/02/18/1274/proyeksi-penduduk-menurut-provinsi-2010%2D%2D-2035.html

Badan Pusat Statistik (2018a) Number of foreign tourist visits to Indonesia by nationality (people), 2018–2019. https://www.bps.go.id/indicator/16/1821/1/number-of-foreign-tourist-visits-to-indonesia-by-nationality.html

Badan Pusat Statistik (2018b) Total Foreign Exchange of Tourism Sector (Billion US $), 2016–2018. https://www.bps.go.id/indicator/16/1160/1/total-foreign-exchange-of-tourism-sector.html

Badan Pusat Statistik (2020) Number of foreign tourist visits per month to Indonesia according to the entrance, 2017-now (visit), *2020*. https://www.bps.go.id/indicator/16/1150/1/number-of-foreign-tourist-visits-per-month-to-indonesia-according-to-the-entrance-2017%2D%2D-now.html

Balasubramaniam KN, Sueur C, Huffman MA, MacIntosh AJ (2020) Primate infectious disease ecology: insights and future directions at the human-macaque interface. In: The behavioral ecology of the Tibetan macaque. Springer, Cham, pp 249–284

Ballantyne R, Packer J, Hughes K, Dierking L (2007) Conservation learning in wildlife tourism settings: lessons from research in zoos and aquariums. Environ Educ Res 13(3):367–383. https://doi.org/10.1080/13504620701430604

Behie AM, Pavelka MSM, Chapman CA (2010) Sources of variation in fecal cortisol levels in howler monkeys in Belize. Am J Primatol 72(7):600–606. https://doi.org/10.1002/ajp.20813

Beisner BA, Heagerty A, Seil SK, Balasubramaniam KN, Atwill ER, Gupta BK, Tyagi PC, Chauhan NPS, Bonal BS, Sinha PR, McCowan B (2015) Human-wildlife conflict: proximate predictors of aggression between humans and rhesus macaques in India. Am J Phys Anthropol 156(2):286–294. https://doi.org/10.1002/ajpa.22649

Bejder L, Samuels A, Whitehead H, Finn H, Allen S (2009) Impact assessment research: use and misuse of habituation, sensitisation and tolerance in describing wildlife responses to anthropogenic stimuli. Mar Ecol Prog Ser 395:177–185. https://doi.org/10.3354/meps07979

Bertolani P, Boesch C (2008) Habituation of wild chimpanzees (*pan troglodytes*) of the south Group at Taï Forest, Côte d'Ivoire: empirical measure of progress. Folia Primatol 79(3):162–171. https://doi.org/10.1159/000111720

Birdlife International (2020) Endemic Bird Areas factsheet: Sulawesi. http://datazone.birdlife.org/eba/factsheet/167

Brotcorne F, Giraud G, Gunst N, Fuentes A, Wandia IN, Beudels-Jamar RC, Poncin P, Huynen M-C, Leca J-B (2017) Intergroup variation in robbing and bartering by long-tailed macaques at Uluwatu Temple (Bali, Indonesia). Primates 58(4):505–516. https://doi.org/10.1007/s10329-017-0611-1

Buckley R (2004) Impacts of ecotourism on terrestrial wildlife. In: Buckley R (ed) Environmental impacts of ecotourism. CABI Publishing, pp 211–228. https://www.cabdirect.org/cabdirect/abstract/20043135899

Büscher B, Sullivan S, Neves K, Igoe J, Brockington D (2012) Towards a synthesized critique of neoliberal biodiversity conservation. Capital Nat Social 23(2):4–30

Butarbutar R, Soemarno S (2013) Environmental effects of ecotourism in Indonesia. J Ind Tour Dev Stud 1(3):97–107

Cai J, Jiang Z, Zeng Y, Li C, Bravery BD (2008) Factors affecting crop damage by wild boar and methods of mitigation in a giant panda reserve. Eur J Wildl Res 54(4):723–728. https://doi.org/10.1007/s10344-008-0203-x

Carne C, Semple S, MacLarnon A, Majolo B, Maréchal L (2017) Implications of tourist–macaque interactions for disease transmission. EcoHealth 14(4):704–717

Chen, H., Yao, H., Ruan, X., Wallner, B., Ostner, J., & Xiang, Z. (2020). Tourism may trigger physiologically stress response of a long-term habituated population of golden snub-nosed monkeys. *Current Zoology*, zoaa076. https://doi.org/10.1093/cz/zoaa076

Chong KL (2020) The side effects of mass tourism: the voices of Bali islanders. Asia Pac J Tour Res 25(2):157–169. https://doi.org/10.1080/10941665.2019.1683591

Chua L (2018) Too cute to cuddle? "Witnessing publics" and interspecies relations on the social media-scape of orangutan conservation. Anthropol Q 91(3):873–903

Chua L, Harrison ME, Fair H, Milne S, Palmer A, Rubis J et al (2020) Conservation and the social sciences: beyond critique and co-optation. A case study from orangutan conservation. People and Nature 2(1):42–60

Clements R, Sodhi NS, Schilthuizen M, Ng PK (2006) Limestone karsts of Southeast Asia: imperiled arks of biodiversity. Bioscience 56(9):733–742

Cochrane J (2006) Indonesian national parks: understanding leisure users. Ann Tour Res 33(4):979–997

Collins EF (2007) Indonesia betrayed: how development fails. University of Hawaii Press

Curtin S (2010) What makes for memorable wildlife encounters? Revelations from "serious" wildlife tourists. J Ecotour 9:149–168

Davis G (1976) Parigi: a social history of the Balinese movement to Central Sulawesi, 1907–1974. Stanford University

Desmond J (1999) Staging tourism: bodies on display from Waikiki to sea world. University of Chicago Press

DeVantier L, Turak E (2004) Managing tourism in Bunaken National Marine park and adjacent waters, North Sulawesi. Natural Resources Management Program III

Dinarto D, Wanto A, Sebastian LC (2020) COVID-19: Impact on Bintan's tourism sector. https://dr.ntu.edu.sg//handle/10356/137356

Direktorat Jenderak KSDAE (2016) Statistik Derektorat Jenderal KSDAE 2016. Kementerian Lingkungan Hidup dan Kehutanan. http://ksdae.menlhk.go.id/assets/publikasi/Draft_final_Statistik_Ditjen_KSDAE_2016_CETAK_FIX.compressed_.pdf

Divinagracia LA, Divinagracia MRG, Divinagracia DG (2012) Digital media-induced tourism: the case of nature-based tourism (NBT) at East Java, Indonesia. Procedia Soc Behav Sci 57:85–94. https://doi.org/10.1016/j.sbspro.2012.09.1161

Duffy R (2002) Trip too far: ecotourism, politics, and exploitation. Earthscan, London

Duffy R (2013) The international political economy of tourism and the neoliberalisation of nature: challenges posed by selling close interactions with animals. Rev Int Polit Econ 20(3):605–626

Dunbar RIM (1987) Demography and reproduction. In: Smuts BB, Bearder SK (eds) Primate societies. University of Chicago Press, pp 240–249

Engel GA, Engel LJ, Schillaci MS, Suaryana KG, Putra A, Fuentes A, Henkel R (2002) Human exposure to herpesvirus B-seropositive macaques, Bali, Indonesia. Emerg Infect Dis 8(8):789

Eshun G, Tonto JNP (2014) Community-based ecotourism: its socio-economic impacts at Boabeng-Fiema monkey sanctuary, Ghana. Bull Geogr Socio-Econ Ser 26(26):67–81. https://doi.org/10.2478/bog-2014-0045

Fedigan LM (2010) Ethical issues faced by field primatologists: asking the relevant questions. Am J Primatol 72(9):754–771. https://doi.org/10.1002/ajp.20814

Fletcher R (2009) Ecotourism discourse: challenging the stakeholders theory. J Ecotour 8(3):269–285

Fletcher R, Neves K (2012) Contradictions in tourism: the promise and pitfalls of ecotourism as a manifold capitalist fix. Environ Soc 3(1):60–77

Fooden J (1969) Taxonomy and evolution of the monkeys of Celebes (primates: Cercopithecidae). Bibl Primatol no. 10. Karger, Basel

Forclime (Forest and Climate Change Programme) (2017) Pengelolaan kawasan konservasi di Indonesia: Pengelolaan saat ini, pembelajaran dan rekomendasi. Deutsche Gesellschaft für Internationale Zusammenarbeit

Freund CA, Achmad M, Kanisius P, Naruri R, Tang E, Knott CD (2020) Conserving orangutans one classroom at a time: evaluating the effectiveness of a wildlife education program for school-aged children in Indonesia. Anim Conserv 23(1):18–27. https://doi.org/10.1111/acv.12513

Fuentes A (2006) Human culture and monkey behavior: assessing the contexts of potential pathogen transmission between macaques and humans. Am J Primatol 68(9):880–896

Fuentes A (2010) Natural cultural encounters in Bali: monkeys, temples, tourists, and ethnoprimatology. Cult Anthropol 25(4):600–624

Fuentes A, Gamerl S (2005) Disproportionate participation by age/sex classes in aggressive interactions between long-tailed macaques (Macaca fascicularis) and human tourists at Padangtegal monkey forest, Bali, Indonesia. Am J Primatol 66(2):197–204

Fuentes A, Hockings KJ (2010) The ethnoprimatological approach in primatology. Am J Primatol 72(10):841–847

Fuentes A, Shaw E, Cortes J (2007) Qualitative assessment of macaque tourist sites in Padangtegal, Bali, Indonesia, and the upper rock nature reserve, Gibraltar. Int J Primatol 28(5):1143–1158

Fuentes A, Riley EP, Dore KM (2017) Ethnoprimatology matters: integration, innovation, and intellectual generosity. In: Dore KM, Riley EP, Fuentes A (eds) Ethnoprimatology: a practical guide to research at the human-nonhuman interface. Cambridge, pp 297–301

Galley G, Clifton J (2004) The motivational and demographic characteristics of research ecotourists: operation Wallacea volunteers in Southeast Sulawesi, Indonesia. J Ecotour 3(1):69–82. https://doi.org/10.1080/14724040408668150

Germani L (2016) Female ano-genital swelling as a complex sexual signal: morphological, behavioral, and hormonal correlates in wild Macaca maura, Rome Tre University

Green, V. M., & Gabriel, K. I. (2020). Researchers' ethical concerns regarding habituating wild-nonhuman primates and perceived ethical duties to their subjects: Results of an online survey. American Journal of Primatology, 82(9). https://doi.org/10.1002/ajp.23178

Griggio C (2015) Looking for experience at Vittangi Moose Park in Swedish Lapland. Scand J Hosp Tour 15(3):244–265. https://doi.org/10.1080/15022250.2014.999015

Grossberg R, Treves A, Naughton-Treves L (2003) The incidental ecotourist: measuring visitor impacts on endangered howler monkeys at a Belizean archaeological site. Environ Conserv 30(01):40–51

Groves C, Shekelle M (2010) The genera and species of Tarsiidae. Int J Primatol 31(6):1071–1082. https://doi.org/10.1007/s10764-010-9443-1

Hakim L, Soemarno M, Hong S-K (2012) Challenges for conserving biodiversity and developing sustainable island tourism in North Sulawesi Province, Indonesia. J Ecol Environ 35(2):61–71

Hansen MF, Ellegaard S, Moeller MM, van Beest FM, Fuentes A, Nawangsari VA, Groendahl C, Frederiksen ML, Stelvig M, Schmidt NM, Traeholt C, Dabelsteen T (2020) Comparative home range size and habitat selection in provisioned and non-provisioned long-tailed macaques (Macaca fascicularis) in Baluran National Park, East Java, Indonesia. Contrib Zool 89(4):393–411. https://doi.org/10.1163/18759866-bja10006

Hanson KT (2017) Primates watching primates watching primates: An ethnoprimatological account of the habituation process in moor macaques (Macaca maura) [Thesis]. San Diego State University

Hanson KT, Riley EP (2017) Beyond neutrality: the human–primate interface during the habituation process. Int J Primatol 1–26. https://doi.org/10.1007/s10764-017-0009-3

Hasyim M (2019) Foreign tourists' perceptions of Toraja as a cultural site in South Sulawesi, Indonesia. Afr J Hosp Tour Leis 8(3):1–13

Hausmann A, Toivonen T, Slotow R, Tenkanen H, Moilanen A, Heikinheimo V, Minin ED (2018) Social media data can be used to understand tourists' preferences for nature-based experiences in protected areas. Conserv Lett 11(1):e12343. https://doi.org/10.1111/conl.12343

Hayward, J., & Rothenberg, M. (2004). Measuring Success in the "Congo Gorilla Forest" Conservation Exhibition. Curator: The Museum Journal, 47(3), 261–282. https://doi.org/10.1111/j.2151-6952.2004.tb00125.x

Higham JES, Shelton EJ (2011) Tourism and wildlife habituation: reduced population fitness or cessation of impact? Tour Manag 32(6):1290–1298

Higuchi Y, Yamanaka Y (2017) Knowledge sharing between academic researchers and tourism practitioners: a Japanese study of the practical value of embeddedness, trust and co-creation. J Sustain Tour 25(10):1456–1473

Hill CM (2000) Conflict of interest between people and baboons: crop raiding in Uganda. Int J Primatol 21(2):299–315. https://doi.org/10.1023/A:1005481605637

Hill CM (2018) Crop foraging, crop losses, and crop raiding. Annu Rev Anthropol 47(1):377–394. https://doi.org/10.1146/annurev-anthro-102317-050022

Hockings KJ, McLennan MR (2012) From forest to farm: systematic review of cultivar feeding by chimpanzees - management implications for wildlife in anthropogenic landscapes. PLoS One 7(4):e33391. https://doi.org/10.1371/journal.pone.0033391

Homsy J (1999) Ape tourism and human diseases: how close should we get? A critical review of the rules and regulations governing park management and tourism for the wild mountain gorilla (Gorilla gorilla beringei). International Gorilla Conservation Programme

Hsu MJ, Kao C-C, Agoramoorthy G (2009) Interactions between visitors and Formosan macaques (*Macaca cyclopis*) at Shou-Shan Nature Park, Taiwan. Am J Primatol 71(3):214–222. https://doi.org/10.1002/ajp.20638

Huda N, Muslikh M, Rini N, Hidayat S (2020) South Sulawesi halal tourism a strategic approach. Jurnal Organisasi dan Manajemen 143:116–120. https://doi.org/10.2991/aebmr.k.200522.024

Hughes K, Packer J, Ballantyne R (2011) Using post-visit action resources to support family conservation learning following a wildlife tourism experience. Environ Educ Res 17(3):307–328. https://doi.org/10.1080/13504622.2010.540644

Hvenegaard GT (2014) Economic aspects of primate tourism associated with primate conservation. In: Russon AE, Wallis J (eds) Primate tourism. Cambridge University Press, pp 259–277

Ilham K, Rizaldi N, Tsuji Y (2018) Effect of provisioning on the temporal variation in the activity budget of urban long-tailed macaques (Macaca fascicularis) in West Sumatra, Indonesia. Folia Primatol 89(5):347–356. https://doi.org/10.1159/000491790

IUCN (2008) 2008 red list of threatened species. IUCN. www.iucnredlist.org

Jacobson SK (2010) Effective primate conservation education: gaps and opportunities. Am J Primatol 72(5):414–419. https://doi.org/10.1002/ajp.20792

Johnston RJ (1998) Exogenous factors and visitor behavior: a regression analysis of exhibit viewing time. Environ Behav 30(3):322–347. https://doi.org/10.1177/001391659803000304

Jones D (2011) An appetite for connection: why we need to understand the effect and value of feeding wild birds. Emu 111(2):i–vii. https://doi.org/10.1071/MUv111n2_ED

Jones-Engel L, Engel GA, Schillaci MA, Rompis A, Putra A, Suaryana KG, Fuentes A, Beer B, Hicks S, White R, Wilson B, Allan JS (2005) Primate-to-human retroviral transmission in Asia. Emerg Infect Dis 11(7):1028–1035

Junaid I (2014) Opportunities and challenges of cultural heritage tourism: Socio-economic politics of sustainable tourism in South Sulawesi province, Indonesia [Thesis, University of Waikato]. https://researchcommons.waikato.ac.nz/handle/10289/8781

Kaburu SSK, Marty PR, Beisner B, Balasubramaniam KN, Bliss-Moreau E, Kaur K, Mohan L, McCowan B (2019) Rates of human–macaque interactions affect grooming behavior among urban-dwelling rhesus macaques (Macaca mulatta). Am J Phys Anthropol 168(1):92–103. https://doi.org/10.1002/ajpa.23722

Kadir AW, Purwanto RH, Poedjirahajoe E (2013) Analisis Stakeholder Pengelolaan Taman Nasional Bantimurung Bulusaraung, Provinsi Sulawesi Selatan (Stakeholder Analysis of Bantimurung Bulusaraung National Park Management, South Sulawesi Province). Jurnal Manusia Dan Lingkungan 20(1):11–21

Keane A, Ramarolahy AA, Jones JPG, Milner-Gulland EJ (2011) Evidence for the effects of environmental engagement and education on knowledge of wildlife laws in Madagascar. Conserv Lett 4(1):55–63. https://doi.org/10.1111/j.1755-263X.2010.00144.x

Kellert SR, Black M, Rush CR, Bath AJ (1996) Human culture and large carnivore conservation in North America. Conserv Biol 10(4):977–990. https://doi.org/10.1046/j.1523-1739.1996.10040977.x

Kemp S (2020) Digital use around the world in July 2020. We Are Social. https://wearesocial.com/us/blog/2020/07/digital-use-around-the-world-in-july-2020

Kinnaird MF, O'Brien TG (1996) Ecotourism in the Tangkoko DuaSudara nature reserve: opening pandora's box? Oryx 30(1):65–73. https://doi.org/10.1017/S0030605300021402

Kling KJ, Hopkins ME (2015) Are we making the grade? Practices and reported efficacy measures of primate conservation education programs. Am J Primatol 77(4):434–448. https://doi.org/10.1002/ajp.22359

Knight J (2009) Making wildlife viewable: habituation and attraction. Society & Animals 17(2):167–184

Knight, J. (2011). Herding monkeys to paradise: How macaque troops are managed for tourism in Japan. Brill.

Kodir A, Tanjung A, Astina IK, Nurwan MA, Nusantara AG, Ahmad R (2020) The dinamics of access on tourism development in Labuan Bajo, Indonesia. GeoJournal Tour Geosit 29(2):662–671. 10.30892/gtg.29222-497

Kurita H (2014) Provisioning and tourism in free-ranging Japanese macaques. In: Russon AE, Wallis J (eds) Primate tourism: a tool for conservation? Cambridge University Press, pp 44–55

Kurniawan Y, Habsari SK, Nurhaeni IDA (2017) Selfie culture: investigating the patterns and various expressions of dangerous selfies and the possibility of government's intervention. Proc J Gov Polit Int Conf 2:324–332

Lappan S, Malaivijitnond S, Radhakrishna S, Riley EP, Ruppert N (2020) The human–primate interface in the new normal: challenges and opportunities for primatologists in the COVID-19 era and beyond. Am J Primatol 82(8):e23176. https://doi.org/10.1002/ajp.23176

Larson CL, Reed SE, Merenlender AM, Crooks KR (2016) Effects of recreation on animals revealed as widespread through a global systematic review. PLoS One 11(12):e0167259. https://doi.org/10.1371/journal.pone.0167259

Lewis MP (ed) (2009) Ethnologue: languages of the world, 16th edn. SIL International

Lowe C (2004) Making the monkey: how the Togean macaque went from "new form" to "endemic species" in Indonesians' conservation biology. Cult Anthropol 19(4):491–516. https://doi.org/10.1525/can.2004.19.4.491

Lowe C (2006) Wild profusion: biodiversity conservation in an Indonesian archipelago. Princeton University Press

MacArthur RH, Wilson EO (1967) The theory of island biogeography. Princeton University Press

Madden F (2006) Gorillas in the garden: human–wildlife conflict at Bwindi impenetrable National Park. Policy Matters 14:180–190

Maréchal L, Semple S, Majolo B, Qarro M, Heistermann M, MacLarnon A (2011) Impacts of tourism on anxiety and physiological stress levels in wild male barbary macaques. Biol Conserv 144(9):2188–2193

Maréchal L, MacLarnon A, Majolo B, Semple S (2016) Primates' behavioural responses to tourists: evidence for a trade-off between potential risks and benefits. Sci Rep 6(1):32465. https://doi.org/10.1038/srep32465

Maryono M, Effendi H, Krisanti M (2019) Tourism carrying capacity to support beach management at Tanjung Bira, Indonesia. Jurnal Segara 15(2):119–126. https://doi.org/10.15578/segara.v15i2.6790

Matsumura S (1991) A preliminary report on the ecology and social behavior of moor macaques (Macaca maurus) in Sulawesi, Indonesia. Kyoto Univ Overseas Res Rep Stud Asian Non-Human Primates 8:27–41

Matsumura S (1998) Relaxed dominance relations among female moor macaques (Macaca maurus) in their natural habitat, South Sulawesi, Indonesia. Folia Primatol 69(6):346–356

McCarthy MS, Matheson MD, Lester JD, Sheeran LK, Li J-H, Wagner RS (2009) Sequences of Tibetan macaque (Macaca thibetana) and tourist behaviors at Mt. Huangshan, China. Prim Conserv 24:145–151

McDougall P (2012) Is passive observation of habituated animals truly passive? J Ethol 30(2):219–223. https://doi.org/10.1007/s10164-011-0313-x

McGowan J, Beaumont LJ, Smith RJ, Chauvenet ALM, Harcourt R, Atkinson SC, Mittermeier JC, Esperon-Rodriguez M, Baumgartner JB, Beattie A, Dudaniec RY, Grenyer R, Nipperess DA, Stow A, Possingham HP (2020) Conservation prioritization can resolve the flagship species conundrum. Nat Commun 11(1):994. https://doi.org/10.1038/s41467-020-14554-z

McIntosh D, Wright PA (2017) Emotional processing as an important part of the wildlife viewing experience. J Outdoor Recreat Tour 18:1–9. https://doi.org/10.1016/j.jort.2017.01.004

McLennan MR, Hill CM (2010) Chimpanzee responses to researchers in a disturbed forest–farm mosaic at Bulindi, western Uganda. Am J Primatol 72(10):907–918. https://doi.org/10.1002/ajp.20839

Melfi V (2010) Selamatkan Yaki! Conservation of Sulawesi crested black macaques *Macaca nigra*. In: Gursky S, Supriatna J (eds) Indonesian primates. Springer, pp 343–356. https://doi.org/10.1007/978-1-4419-1560-3_19

Merker S, Driller C, Dahruddin H, Wirdateti S, Perwitasari-Farajallah D, Shekelle M (2010) Tarsius wallacei: a new tarsier species from Central Sulawesi occupies a discontinuous range. Int J Primatol 31(6):1107–1122. https://doi.org/10.1007/s10764-010-9452-0

Moore M, Townsend M, Oldroyd J (2006) Linking human and ecosystem health: the benefits of community involvement in conservation groups. EcoHealth 3(4):255–261. https://doi.org/10.1007/s10393-006-0070-4

Morgan D, Mundry R, Sanz C, Ayina CE, Strindberg S, Lonsdorf E, Kühl HS (2018) African apes coexisting with logging: comparing chimpanzee (*pan troglodytes troglodytes*) and gorilla (*Gorilla gorilla gorilla*) resource needs and responses to forestry activities. Biol Conserv 218:277–286. https://doi.org/10.1016/j.biocon.2017.10.026

Morrow K (2018) Risky business: Causes and conservation implications of human-moor macaque (Macaca maura) interactions in South Sulawesi, Indonesia [PhD Thesis]. San Diego State University

Morrow KS, Glanz H, Ngakan PO, Riley EP (2019) Interactions with humans are jointly influenced by life history stage and social network factors and reduce group cohesion in moor macaques (*Macaca maura*). Sci Rep 9(1):20162. https://doi.org/10.1038/s41598-019-56288-z

Muehlenbein MP (2017) Primates on display: potential disease consequences beyond bushmeat. Am J Phys Anthropol 162:32–43. https://doi.org/10.1002/ajpa.23145

Muehlenbein MP, Martinez LA, Lemke AA, Ambu L, Nathan S, Alsisto S, Sakong R (2010) Unhealthy travelers present challenges to sustainable primate ecotourism. Travel Med Infect Dis 8(3):169–175

Nakamura M, Nishida T (2009) Chimpanzee tourism in relation to the viewing regulations at the Mahale Mountains National Park, Tanzania. Prim Conserv 24(1):85–90. https://doi.org/10.1896/052.024.0106

Narat V, Pennec F, Simmen B, Ngawolo JCB, Krief S (2015) Bonobo habituation in a forest–savanna mosaic habitat: influence of ape species, habitat type, and sociocultural context. Primates 56(4):339–349. https://doi.org/10.1007/s10329-015-0476-0

Nekaris BKA-I, Campbell N, Coggins TG, Rode EJ, Nijman V (2013) Tickled to death: Analysing public perceptions of 'cute' videos of threatened species (slow lorises – *Nycticebus* spp.) on web 2.0 sites. PLoS One 8(7):e69215. https://doi.org/10.1371/journal.pone.0069215

Nijman V, Nekaris KA-I (2010) Testing a model for predicting primate crop-raiding using crop- and farm-specific risk values. Appl Anim Behav Sci 127(3):125–129. https://doi.org/10.1016/j.applanim.2010.08.009

Nilsson D, Gramotnev G, Baxter G, Butler JRA, Wich SA, McAlpine CA (2016) Community motivations to engage in conservation behavior to conserve the Sumatran orangutan. Conserv Biol 30(4):816–826. https://doi.org/10.1111/cobi.12650

Nur AC, Akib H, Niswaty R, Aslinda A, Zaenal H 2019. *Development partnership strategy tourism destinations integrated and infrastructure in South Sulawesi*, Indonesia, pp 271–283

Okamoto K, Matsumura S, Watanabe K (2000) Life history and demography of wild moor macaques (*Macaca maurus*): summary of ten years of observations. Am J Primatol 52(1):1–11

Olival KJ, Hosseini PR, Zambrana-Torrelio C, Ross N, Bogich TL, Daszak P (2017) Host and viral traits predict zoonotic spillover from mammals. Nature 546(7660):646–650. https://doi.org/10.1038/nature22975

Orams M (1994) Creating effective interpretation for managing interaction between tourists and wildlife. Aust J Environ Educ 10:21–34. https://doi.org/10.1017/S0814062600003062

Osborn FV, Hill CM (2005) Techniques to reduce crop loss: human and technical dimensions in Africa. In: Woodroffe R, Thirgood S, Rabinowitz A (eds) People and wildlife, conflict or co-existence? Cambridge University Press, pp 72–85

Otsuka R, Yamakoshi G (2020) Analyzing the popularity of YouTube videos that violate mountain gorilla tourism regulations. PLoS One 15(5):e0232085. https://doi.org/10.1371/journal.pone.0232085

Padua SM, Dietz LA, Souza MG, Santos GR, Kleiman DG, Rylands AB (2002) In situ conservation education and the lion tamarins. In: Lion tamarins: biology and conservation. Smithsonian Inst Pr, Washington DC, pp 315–335

Pambudi D, McCaughey N, Smyth R (2009) Computable general equilibrium estimates of the impact of the Bali bombing on the Indonesian economy. Tour Manag 30(2):232–239. https://doi.org/10.1016/j.tourman.2008.06.007

Papworth SK, Nghiem TPL, Chimalakonda D, Posa MRC, Wijedasa LS, Bickford D, Carrasco LR (2015) Quantifying the role of online news in linking conservation research to Facebook and twitter. Conserv Biol 29(3):825–833. https://doi.org/10.1111/cobi.12455

Parker GE, Osborn FV (2006) Investigating the potential for chilli *capsicum* spp. to reduce human-wildlife conflict in Zimbabwe. Oryx 40(3):343–346. https://doi.org/10.1017/S0030605306000822

Paulus, A. (2009). Impacts of ecotourism on the behaviour of Sulawesi crested black macaques (Macaca nigra) and spectral tarsiers (Tarsius spectrum) in the Tangkoko-Batuangus Nature Reserve, North Sulawesi, Indonesia [The University of Plymouth]

Pearce J, Moscardo G (2015) In: Hay R (ed) Social representations of tourist selfies: new challenges for sustainable tourism. James Cook University, pp 59–73. http://www.besteducation-network.org/page_wgkE66

Powell RB, Ham SH (2008) Can ecotourism interpretation really lead to pro-conservation knowledge, attitudes and behaviour? Evidence from the Galapagos Islands. J Sustain Tour 16(4):467–489. https://doi.org/10.1080/09669580802154223

Priston NEC, Wyper RM, Lee PC (2012) Buton macaques (*Macaca ochreata brunnescens*): crops, conflict, and behavior on farms. Am J Primatol 74(1):29–36. https://doi.org/10.1002/ajp.21003

Prodjo WA (2017) 10 Destinasi "Bali Baru", 4 Destinasi Jadi Prioritas. KOMPAS.com. https://travel.kompas.com/read/2017/11/18/122700027/10-destinasi-bali-baru-4-destinasi-jadi-prioritas.%20Retrieved%20July%2030

Purvis A, Gittleman JL, Cowlishaw G, Mace GM (2000) Predicting extinction risk in declining species. Proc R Soc Lond Ser B Biol Sci 267(1456):1947–1952. https://doi.org/10.1098/rspb.2000.1234

Putri IASLP (2016) Handicraft of butterflies and moths (Insecta: *Lepidoptera*) in Bantimurung nature Recreation Park and its implications on conservation. Biodiversit J Biol Divers 17(2):Article 2. https://doi.org/10.13057/biodiv/d170260

Putri IASLP (2020) Challenges in initiating *Tarsius fuscus'* creative ecotourism at Bantimurung Bulusaraung National Park. IOP Conf Ser: Earth Environ Sci 533:012005. https://doi.org/10.1088/1755-1315/533/1/012005

Putri IASLP, Ansari F, Susilo A (2020) Response of bird community toward tourism activities in the karst area of Bantimurung Bulusaraung National Park. J Qual Assur Hosp Tour 21(2):146–167. https://doi.org/10.1080/1528008X.2019.1631725

Rahbiah S, Salman D, Yusran IMF (2016) The role of Bantimurung ecotourism for community's livelihood in Maros, province of South Sulawesi, Indonesia. Asian J Appl Sci 4(2):Article 2. https://python.zzx.us/index.php/AJAS/article/view/3611

Rahmat MA, Umar S, Sangadji MN (2016) Potential and strategy of ecotourism management in the Lore Lindu National Park (case study in Sigi regency, Central Sulawesi Province, Indonesia). J Tour Hosp Sport 22:110–121

Ram S, Venkatachalam S, Sinha A (2003) Changing social strategies of wild female bonnet macaques during natural foraging and on provisioning. Current Science, 780–790.

Rasmussen, D. R. (1991). Observer influence on range use of Macaca arctoides after 14 years of observation. *Laboratory Primate Newsletter, 30*(3), 6–11.

Rasmussen RD (1998) Changes in range use of Geoffroy's tamarins (*Saguinus geoffroyi*) associated with habituation to observers. Folia Primatol 69(3):153–159. https://doi.org/10.1159/000021577

Remis MJ, Jost Robinson CA (2020) Elephants, hunters, and others: integrating biological anthropology and multispecies ethnography in a conservation zone. Am Anthropol 122(3):459–472

Reynolds PC, Braithwaite D (2001) Towards a conceptual framework for wildlife tourism. Tour Manag 22(1):31–42. https://doi.org/10.1016/S0261-5177(00)00018-2

Riley EP (2007a) The human–macaque interface: conservation implications of current and future overlap and conflict in Lore Lindu National Park, Sulawesi, Indonesia. Am Anthropol 109(3):473–484

Riley EP (2007b) Flexibility in diet and activity patterns of *Macaca tonkeana* in response to anthropogenic habitat alteration. Int J Primatol 28(1):107–133. https://doi.org/10.1007/s10764-006-9104-6

Riley EP (2010) The endemic seven: four decades of research on the Sulawesi macaques. Evol Anthropol Issues News Reviews 19(1):22–36

Riley EP (2013) Contemporary primatology in anthropology: beyond the epistemological abyss. Am Anthropol 115(3):411–422

Riley, EP (2019) The Promise of Contemporary Primatology. Routledge.

Riley EP, Bezanson M (2018) Ethics of primate fieldwork: toward an ethically engaged primatology. Annu Rev Anthropol 47(1):493–512. https://doi.org/10.1146/annurev-anthro-102317-045913

Riley EP, Sagnotti C, Carosi M, Oka NP (2014) Socially tolerant relationships among wild male moor macaques (*Macaca maura*). Behaviour 151(7):1021–1044. https://doi.org/10.1163/1568539X-00003182

Riley CM, DuVall-Lash AS, Jayasri SL, Koenig BL, Klegarth AR, Gumert MD (2016) How living near humans affects Singapore's urban macaques. In: Waller MT (ed) Ethnoprimatology: primate conservation in the 21st century. Springer International Publishing, pp 283–300. https://doi.org/10.1007/978-3-319-30469-4_16

Roberge J-M, Angelstam P (2004) Usefulness of the Umbrella Species Concept as a Conservation Tool. Conservation Biology, 18(1), 76–85. https://doi.org/10.1111/j.1523-1739.2004.00450.x

Rodger K, Calver M (2005) Natural science and wildlife tourism. In: Newsome D, Moore SA, Dowling RK (eds) Wildlife tourism. Channel View Publications, Clevedon, UK, pp 217–234

Rodger K, Moore SA, Newsome D (2010) Wildlife tourism science and scientists: barriers and opportunities. Soc Nat Res Int J 23(8):679–694

Rodriguez-Lopez M, Mignucci-Giannoni A (1999) Mortality of a friendly wild roughtooth dolphin (Steno bredanensis) in Aruba. In Proceedings of the 13th Biennial Conference of Marine Mammals, Wailea, Hawaii

Ross C (1992) Life history patterns and ecology of macaque species. Primates 33(2):207–215

Ross S, Wall G (1999) Evaluating ecotourism: the case of North Sulawesi, Indonesia. Tour Manag 20(6):673–682. https://doi.org/10.1016/S0261-5177(99)00040-0

Ross SR, Lukas KE, Lonsdorf EV, Stoinski TS, Hare B, Shumaker R, Goodall J (2008) Inappropriate use and portrayal of chimpanzees. Science 319(5869):1487–1487

Ross SR, Vreeman VM, Lonsdorf EV (2011) Specific image characteristics influence attitudes about chimpanzee conservation and use as pets. PLoS One 6(7):e22050

Rusdianto E (2020, May 10) Pandemi Corona, Kondisi Macaca di Maros Makin Rawan. Mongabay Environmental News Indonesia. https://www.mongabay.co.id/2020/05/10/pandemi-corona-kondisi-macaca-di-maros-makin-rawan/

Russon AE, Susilo A (2014) Orangutan tourism and conservation: 35 years' experience. In: Primate tourism: a tool for conservation? Cambridge University Press, pp 76–97

Russon AE, Wallis J (2014) Primate tourism: a tool for conservation? Cambridge University Press

Ryan SJ, Walsh PD (2011) Consequences of non-intervention for infectious disease in African great apes. PLoS One 6(12):e29030. https://doi.org/10.1371/journal.pone.0029030

Sagnotti C (2013) Diet preference and habitat use in relation to reproductive states in females of a wild group of Macaca maura inhabiting Karaenta forest in South Sulawesi. Hasanuddin University

Sahabuddin W (2019) Lakeside resort based on eco-architecture. IOP Conf Ser Mater Sci Eng 471:082073. https://doi.org/10.1088/1757-899X/471/8/082073

Salas L, Dickman C, Helgen K, Flannery T (2019) Ailurops ursinus. The IUCN Red List of Threatened Species.

Santika T, Wilson KA, Budiharta S, Law EA, Poh TM, Ancrenaz M, Struebig MJ, Meijaard E (2019) Does oil palm agriculture help alleviate poverty? A multidimensional counterfactual assessment of oil palm development in Indonesia. World Dev 120:105–117. https://doi.org/10.1016/j.worlddev.2019.04.012

Sapolsky RM (2014) Some pathogenic consequences of tourism for non-human primates. In: Russon AE, Wallis J (eds) Primate tourism: a tool for conservation. Cambridge University Press, pp 147–155

Scarduelli P (2005) Dynamics of cultural change among the Toraja of Sulawesi. The commoditization of tradition. Anthropos 100(2):389–400

Sengupta A, Radhakrishna S (2018) The hand that feeds the monkey: mutual influence of humans and rhesus macaques (*Macaca mulatta*) in the context of provisioning. Int J Primatol 39(5):817–830. https://doi.org/10.1007/s10764-018-0014-1

Sengupta A, Radhakrishna S (2020) Factors predicting provisioning of macaques by humans at tourist sites. Int J Primatol 41(3):471–485. https://doi.org/10.1007/s10764-020-00148-5

Sengupta A, McConkey KR, Radhakrishna S (2015) Primates, provisioning and plants: impacts of human cultural behaviours on primate ecological functions. PLoS One 10(11):e0140961. https://doi.org/10.1371/journal.pone.0140961

Setchell JM, Fairet E, Shutt K, Waters S, Bell S (2017) Biosocial conservation: integrating biological and ethnographic methods to study human–primate interactions. Int J Primatol 38(2):401–426

Shannon G, Larson CL, Reed SE, Crooks KR, Angeloni LM (2017) Ecological consequences of ecotourism for wildlife populations and communities. In: Blumstein DT, Geffroy B, Samia DSM, Bessa E (eds) Ecotourism's promise and peril: a biological evaluation. Springer International Publishing, pp 29–46. https://doi.org/10.1007/978-3-319-58331-0_3

Shekelle M, Groves CP, Maryanto I, Mittermeier RA (2017) Two new tarsier species (*Tarsiidae*, primates) and the biogeography of Sulawesi, Indonesia. Prim Conserv 31:61–69

Shekelle M, Groves CP, Maryanto I, Mittermeier RA, Salim A, Springer MS (2019) A new tarsier species from the Togean Islands of Central Sulawesi, Indonesia, with references to Wallacea and conservation on Sulawesi. Prim Conserv 33:65–73

Sherrow HM (2010) Conservation education and primates: twenty-first century challenges and opportunities. Am J Primatol 72(5):420–424. https://doi.org/10.1002/ajp.20788

Shettel-Neuber J (1988) Second and third-generation zoo exhibits: a comparison of visitor, staff, and animal responses. Environ Behav 20(4):452–473. https://doi.org/10.1177/0013916588204005

Shutt K, Heistermann M, Kasim A, Todd A, Kalousova B, Profosouva I, Petrzelkova K, Fuh T, Dicky J-F, Bopalanzognako J-B, Setchell JM (2014) Effects of habituation, research and ecotourism on faecal glucocorticoid metabolites in wild western lowland gorillas: implications for conservation management. Biol Conserv 172:72–79. https://doi.org/10.1016/j.biocon.2014.02.014

Sinha A, Mukhopadhyay K, Datta-Roy A, Ram S (2005) Ecology proposes, behaviour disposes: ecological variability in social organization and male behavioural strategies among wild bonnet macaques. Curr Sci 89(7):1166–1179

Smith GK (2012) Water fun park and tourist caves, amidst tropical tower karst. ACKMA Journal 86:5

Stem C, Margoluis R, Salafsky N, Brown M (2005) Monitoring and evaluation in conservation: a review of trends and approaches. Conserv Biol 19(2):295–309. https://doi.org/10.1111/j.1523-1739.2005.00594.x

Stronza A (2007) The economic promise of ecotourism for conservation. J Ecotour 6(3):210–230

Stronza A, Pêgas F (2008) Ecotourism and conservation: two cases from Brazil and Peru. Hum Dimens Wildl 13(4):263–279. https://doi.org/10.1080/10871200802187097

Stronza AL, Hunt CA, Fitzgerald L A (2019) Ecotourism for Conservation? 27.

Supriatna J, Shekelle M, Fuad HAH, Winarni NL, Dwiyahreni AA, Farid M, Mariati S, Margules C, Prakoso B, Zakaria Z (2020) Deforestation on the Indonesian island of Sulawesi and the loss of primate habitat. Glob Ecol Conserv 24:e01205. https://doi.org/10.1016/j.gecco.2020.e01205

Suriamihardja D (2010) Sustainable tourism: The case in South Sulawesi, pp 1–12. https://www.researchgate.net/profile/Dadang_Suriamihardja/publication/271964727_Sustainable_Tourism_the_Case_in_South_Sulawesi/links/54d7954c0cf25013d03aae1a/Sustainable-Tourism-the-Case-in-South-Sulawesi.pdf

Tam C-L (2019) Branding Wakatobi: marine development and legitimation by science. Ecol Soc 24(3):23. https://doi.org/10.5751/ES-11095-240323

Tangian D, Djokosetiyanto D, Kholil K, Munandar A (2015) Model of ecotourism management in small islands of Bunaken National Park, North Sulawesi. J Ind Tour Dev Stud 3(2):75–84

Tenkanen H, Di Minin E, Heikinheimo V, Hausmann A, Herbst M, Kajala L, Toivonen T (2017) Instagram, Flickr, or twitter: assessing the usability of social media data for visitor monitoring in protected areas. Sci Rep 7(1):17615. https://doi.org/10.1038/s41598-017-18007-4

Trolliet F, Serckx A, Forget P-M, Beudels-Jamar RC, Huynen M-C, Hambuckers A (2016) Ecosystem services provided by a large endangered primate in a forest-savanna mosaic landscape. Biol Conserv 203:55–66. https://doi.org/10.1016/j.biocon.2016.08.025

Tsing AL (2005) Friction: an ethnography of global connection. Princeton University Press. https://books.google.com/books?hl=en & lr= & id=pCwEA1A_XPcC & oi=fnd & pg=PR2 & dq=Friction:+An+Ethnography+of+Global+Connection & ots=c0eu0e1zYm & sig=Ci_qx1I5wswuOcSpT6QBDuo6R0E

Tsuji Y, Su H-H (2018) Macaques as seed dispersal agents in Asian forests: a review. Int J Primatol 39(3):356–376. https://doi.org/10.1007/s10764-018-0045-7

Tutin CEG, Fernandez M (1991) Responses of wild chimpanzees and gorillas to the arrival of primatologists: behaviour observed during habituation. In: Box HO (ed) Primate responses to environmental change. Springer, Netherlands, pp 187–197. https://doi.org/10.1007/978-94-011-3110-0_10

van den Bergh GD, Li B, Brumm A, Grün R, Yurnaldi D, Moore MW, Kurniawan I, Setiawan R, Aziz F, Roberts RG, Suyono S, M., Setiabudi, E., & Morwood, M. J. (2016) Earliest hominin occupation of Sulawesi, Indonesia. Nature 529:208–211

von Heland F, Clifton J (2015) Whose threat counts? Conservation narratives in the Wakatobi National Park, Indonesia. Conserv Soc 13(2):154–165

Wakka AK, Muin N, Purwanti R (2015) Toward collaborative management of Bantimurung Bulusaurang National Park, South Sulawesi Province. Journal Penelitian Kehutanan Wallacea 4(1):41–50. https://doi.org/10.18330/jwallacea.2015.vol4iss1pp41-50

Wallis J, Lee RD (1999) Primate conservation: the prevention of disease transmission. Int J Primatol 20(6):803–826. https://doi.org/10.1023/A:1020879700286

Waluyo H, Sadikin SR, Gustami WP (2005) An economic valuation of biodiversity in the karst area of Maros, South Sulawesi, Indonesia. Biodiversity 6(2):24–26

Watanabe K, Matsumura S (1996) Social organization of moor macaques (*Macaca maurus*) in the Karaenta nature reserve, South Sulawesi, Indonesia. In: Variations in the Asian macaques. Tokai University Press, pp 147–162

Waters S, Setchell JM, Maréchal L, Oram F, Wallis J, Cheyne SM (2021) Best Practice Guidelines for Responsible Images of Non-Human Primates. IUCN/SSC Specialist Group.

Waylen K (2010). The implications of local views and institutions for the outcomes of community-based conservation [University of London]. https://www.iccs.org.uk/wp-content/thesis/PhD-KerryWaylen2010.pdf

Waylen KA, McGowan PJK, Group PS, Milner-Gulland EJ (2009) Ecotourism positively affects awareness and attitudes but not conservation behaviours: a case study at Grande Riviere, Trinidad. Oryx 43(3):343–351. https://doi.org/10.1017/S0030605309000064

Welker M (2014) Enacting the corporation: an American mining firm in post-authoritarian Indonesia. University of California Press

West P (2006) Conservation is our government now: the politics of ecology in Papua New Guinea. Duke University Press, Durham, NC

West P, Carrier JG (2004) Getting away from it all? Ecotourism and authenticity. Curr Anthropol 45(4):483–498

Westin JL (2017) Habituation to tourists: protective or harmful? In: Dore KM, Riley EP, Fuentes A (eds) Ethnoprimatology: a practical guide to research at the human-nonhuman interface. Cambridge, pp 15–28

Whitten T, Mustafa M, Henderson GS (2002) The ecology of Sulawesi. Periplus

Wilcove D (1993) Getting ahead of the extinction curve. Ecol Appl 3(2):218–220. https://doi.org/10.2307/1941824

Williamson EA, Fawcett K (2008) Long-term research and conservation of the Virunga mountain gorillas. In: Wrangham R, Ross E (eds) Science and conservation in African forests: the benefits of long-term research. Cambridge University Press, Cambridge, pp 213–229

Williamson EA, Feistner A (2011) Habituating primates: processes, techniques, variables and ethics. In: Setchell JM, Curtis DJ (eds) Field and laboratory methods in primatology: a practical guide, 2nd edn. Cambridge University Press, pp 33–49

Williamson EA, Macfie EJ (2014) Guidelines for best practice in great ape tourism. In: Russon AE, Wallis J (eds) Primate tourism. Cambridge University Press, pp 292–310

Wright PC, Andriamihaja B, King SJ, Guerriero J, Hubbard J (2014) Lemurs and tourism in Ranomafana National Park, Madagascar: economic boom and other consequences. In: Russon AE, Wallis J (eds) Primate tourism: a tool for conservation? Cambridge University Press

Yamashita S (1994) Manipulating ethnic tradition: the funeral ceremony, tourism, and television among the Toraja of Sulawesi. Indonesia 58:69–82. https://doi.org/10.2307/3351103

Yu H, Zhang M, Yang J, Yao H, Tian Y (2011) A discussion on ecotourism and exploitation of Chinese primates—the case of the golden monkey ecotourism project in Shennongjia nature reserve. Econ Res Guid 126(16):141–144

Zak A (2016) Mischievous monkeys: Ecological and ethnographic component of crop raiding by moor macaques (Macaca maura) in South Sulawesi, Indonesia [San Diego State University]. https://sdsu-primo.hosted.exlibrisgroup.com/primo-explore/fulldisplay?docid=dedupmrg198957969&context=PC&vid=01CALS_SDL&lang=en_US&search_scope=EVERYTHING&adaptor=primo_central_multiple_fe&tab=everything&query=any,contains,alison%20zak&sortby=rank&offset=0

Zak AA, Riley EP (2017) Comparing the use of camera traps and farmer reports to study crop feeding behavior of moor macaques (Macaca maura). Int J Primatol 38(2):224–242. https://doi.org/10.1007/s10764-016-9945-6

Zeppel H, Muloin S (2008) Aboriginal interpretation in Australian wildlife tourism. J Ecotour 7(2–3):116–136. https://doi.org/10.1080/14724040802140493

Chapter 8
Primates and Primatologists: Reflecting on Two Decades of Primatological and Ethnoprimatological Research, Tourism, and Conservation at the Ubud Monkey Forest

Michaela E. Howells, James E. Loudon, Fany Brotcorne, Jeffrey V. Petterson, I. Nengah Wandia, I. G. A. Arta Putra, and Agustín Fuentes

Abstract The interface between humans and nonhuman primates (NHP) is expanding and intensifying. The Ubud Monkey Forest in Bali, Indonesia is a working example of a multi-use, multi-species interface. This forest is an active religious space, top tourist attraction, critical regional economic contributor, core habitat for long-tailed macaques (Macaca fascicularis), and an important research and training location. This chapter explores two decades of successes, challenges, and ethnoprimatological research at this site. Successes in the forest include developing a durable management system to effectively address the affairs of the forest and surrounding villages, strong educational and outreach programs for the local community, initiatives to reduce plastic pollution and erosion, the establishment of extensive veterinary care, and the development of a provisioning program suited to

M. E. Howells (✉)
Department of Anthropology, University of North Carolina-Wilmington, Wilmington, NC, USA
e-mail: howellsm@uncw.edu

J. E. Loudon
Department of Anthropology, East Carolina University, Greenville, NC, USA

F. Brotcorne
Research Unit SPHERES, Department of Biology, Ecology and Evolution, University of Liège, Liège, Belgium

J. V. Petterson
Department of Anthropology, Ohio State University, Columbus, OH, USA

I. N. Wandia · I. G. A. Arta Putra
Primate Division of Natural Resources and Environment Research Center, Universitas Udayana, Denpasar, Bali, Indonesia

A. Fuentes
Department of Anthropology, Princeton University, Princeton, NJ, USA

© The Author(s), under exclusive license to Springer Nature Switzerland AG 2022
S. L. Gursky et al. (eds.), *Ecotourism and Indonesia's Primates*, Developments in Primatology: Progress and Prospects, https://doi.org/10.1007/978-3-031-14919-1_8

153

the nutritional requirements of the macaques. Challenges to the forest include rapid macaque population growth, need for interventions for aggressive behavior, and expansion of macaque ranging patterns outside of the forest. Research at the site is the result of robust collaborations with Balinese, Canadian, European, and US universities, Indonesian and regional Balinese governments, and local stakeholders working together to understand the dynamics of primate tourism, ethnoprimatology, and human-NHP interconnections. In addition, the forest has been a critical space for field schools, scientific training for developing STEM professionals, and collaborative research. This research and training relies on support from local communities, and cultural competency that includes strong communication, an appreciation of historical context, and an understanding of religious and cultural institutions. Today, the Ubud Monkey Forest acts as an excellent example of how to use tourist revenues to stabilize and improve the health of NHP, conserve forest fragments considered sacred or important to local people, and incorporate monies into local economies to improve the lives and livelihoods of the primates who share these landscapes.

8.1 Introduction

As human populations grow, the intensity of overlap between humans and wildlife increases. Across the globe, animal populations may be culled, harvested for food, or protected and celebrated. The treatment of wildlife is often regulated by formal laws and a society's perceptions and attitudes toward the animal species they live among. In primatology, human-alloprimate relationships are the primary focus of the sub-discipline of ethnoprimatology. Ethnoprimatologists employ the conventional methods and theory of primatology and cultural anthropology in an aim to fully understand human-nonhuman primate (NHP) interconnections from their ecological associations and the symbolic perspectives of the local people (Dore et al. 2017; Fuentes 2012; Riley, 2019).

To date, many ethnoprimatological inquiries have focused on the interplays between humans and macaques (*Macaca* sp.) (e.g., Radhakrishna et al. 2013). The genus *Macaca*, is among the most successful group of primates and is found throughout central and south Asia. Their large geographic distribution is linked to their high degrees of behavioral plasticity, cognitive complexity, and ability to live in a variety of habitats, including anthropogenicallydisturbed forests and cities (Sueur et al. 2011; Gumert et al. 2011; Fuentes 2013).

Globalization and the related rise of tourism exert extensive pressures on human and natural systems. The intensity of overlap between humans and wildlife is ever increasing as human populations continue to grow. In addition to using more land for food production, growing populations also require more space for housing and economic activity. For example, in Bali, Indonesia, there has been a shift toward urbanization and habitat conversion resulting in more land dedicated to domestic and international tourism, reducing the land available for traditional rice agriculture (*sawah*). Increased investment in a tourism economy has resulted in additional protections to some forest fragments associated with Hindu temple complexes. These provide habitat for a culturally and economically important species of monkey, the long-tailed macaque (*Macaca fascicularis*).

Macaques living at these temple complexes figure prominently into the Balinese Hindu philosophy of *Tri Hita Karana* that posits that health, happiness, and prosperity are achieved by living harmoniously with the gods, the spiritual world, and the environment (Fuentes 2010; Lansing et al. 2017). These religious and cultural traditions result in dynamic relationships between the local Balinese people and macaques that have spanned centuries. Human-macaque interfaces also vary based on the management of "monkey forests" and the availability of tourists. Across Bali, the varying degrees of human-macaque overlaps at temple sites and nature reserves provide an excellent model to study the interfaces between social, economic, and environmental spheres.

In this chapter, we will focus on the Ubud Monkey Forest located primarily in the Balinese village Padangtegal and bordering Ubud, and Nyuh Kuning. Although not all research highlighted in this chapter is considered ethnoprimatological, it is conducted in a context of intensive human-nonhuman primate (NHP) interface. This forest is an active religious space, top tourist attraction, and a critical economic contributor to the region. At present, the Ubud Monkey Forest has the highest intensity of human-macaque overlap of any of the 63 macaque sites we have identified across Bali (Fuentes et al. 2005).

8.2 Background History and Overview of the Forest

The Ubud Monkey Forest is a ~ 800-year-old Balinese Hindu temple complex that is used by the Balinese for active worship, rituals, celebrations, and ceremonies, including open air cremations (Fuentes et al. 2011; Fig. 8.1). The first records of the site are noted in inscriptions from 1181 CE and indicate these landscapes were designated as hunting grounds belonging to King Jayapangus. In 1998, the forest-temple complex consisted of 8 hectares of tropical rainforest including 125 plant species (Wijana and Wesnawa 2018). Successive land expansions over the last 20 years have increased the sanctuary to 20.5 hectares in 2020. This includes a large welcome center with a café and parking lot. Today, the site is home to ~1100 fully habituated and provisioned long-tailed macaques (Brotcorne 2014; Giraud et al. in prep).

Many of the remaining forest fragments that are present throughout Bali are associated with Hindu temple complexes. These complexes frequently include a primary temple (*Pura Agung*), and other smaller temples that are enclosed with stone wall and "protected" by stone statues. Within these temple complexes, there are several smaller altars to which the local people make offerings to the gods. Balinese Hindu temple complexes are also utilized by the long-tailed macaques who may be provisioned by the local people due to their associations with the gods as

Fig. 8.1 The Balinese Hindu temple inside the Ubud monkey forest

outlined by the foundations of *Tri Hita Karana* and other important Hindu canons including the *Ramayana* and *Mahabharata* (Wheatley 1999; Fuentes 2010). The temple complex at the Ubud Monkey Forest is comprised of the *Pura Dalem*, the primary and largest temple in the monkey forest, as well as the *Pura Puncak* and *Pura Prajapati*.

8.3 Background and History of Primate-Focused Tourism at the Ubud Monkey Forest

The Ubud Monkey Forest is a major employer in the area. The majority of the ~100 employees by the forest live in the village of Padangtegal. There are a variety of specialists and generalists who are employed through the site. Specialists include administration and management, veterinary and health center workers, and information technology staff. These individuals typically hold university degrees associated with their skills and their job descriptions have a narrower focus. Generalists are equally skilled in their jobs, and include temple guards, members of the conservation team, cleaning crew, security, welcome center, café staff, ticket collection, and parking team members. All members of the team undergo significant ongoing education and training associated with their roles in the forest.

Approximately 40 temple guards oversee tourist-macaque interactions and provision the macaques (Fuentes et al. 2011; Brotcorne 2014). These guards have built long-term relationships with the monkeys. Their interactions tend to be relaxed and familiar. In some cases, guards and monkeys share multiyear familiar relationships with each other where monkeys chose to sit near specific guards and allow themselves to be pet or groomed, played with, and/or playfully teased.

The degree to which macaques are habituated and provisioned varies from temple to temple across the island. At some forest sites (e.g., Alas Nengan, Bukit Gumang, Pulaki, Uluwatu, and Sangeh), local macaques are tolerated when they are in the temple complex, but otherwise may be viewed as crop raiders or nuisances (Loudon et al. 2006). However, at the Ubud Monkey Forest, the macaque population is managed by the temple guards and is generally viewed favorably. Macaques at this site are provisioned with sweet potatoes, corn, papaya leaves, and bananas. Provisioning programs were implemented to increase habituation, reduce crop raiding, provide adequate nutrition, and improve their overall health and coat conditions.

The forest provides direct economic benefits to the village of Padangtegal. Tourist dollars, which are primarily generated from entrance fees, provide critical funding for schools and public health infrastructure. However, the forest leadership recently chose to change the forest's name for marketing purposes and associate their sacred temple with the neighboring, and more widely known, village of Ubud. As such, the name of the site is the Ubud Monkey Forest, but it is also referred to as the Sacred Monkey Forest Sanctuary and Mandala Suci Wenara Wana and formally known as Padangtegal Wenara Wana (Padangtegal Monkey Forest). This rebranding is indicative of the importance of tourism to this community.

In 2018, the forest and village leadership completed an elaborate welcome center and large parking lot. These new structures redirect traffic from the Ubud entrance and concentrate access through the villages of Padangtegal and Nyuh Kuning. These changes were implemented to address the exponential growth in tourists and associated unsustainable traffic and parking pressures. These modifications also acted to redirect tourists through Padangtegal and into their restaurants, shops, and accommodations.

The Ubud Monkey Forest is hemmed in by locally owned tourist shops, sawah fields, and a network of busy roads (Fig. 8.2). Its location at the intersection of three villages offers tourists a range of dining, shopping, and accommodation options from local and international vendors. Alleyways repurposed as local art galleries are frequently found next to international chains including Billabong and Quiksilver. The Gianyar Regency (where Ubud is located) is renowned for its artists and is referred to as the "Arts and Cultural District of Bali" (Vickers 2019). However, many of those artists have had to take on additional employment in the tourism industry due to economic pressures. Ubud was also highlighted in the bestselling memoir *Eat, Pray, Love*, which resulted in a flurry of business ventures that offer all three components of the book (Bell 2019). Tourists can eat their gelato, take a yoga class, and meet with a medicinal healer all before lunch (Fig. 8.3).

The island of Bali has been transformed fundamentally by international tourism with ~4.6 million people visiting the island each year before the spread of COVID-19 (Bali Tourism Office 2017). Prior to WWII, Bali was a popular tourist destination for Japanese citizens. Tourism slowed post war, and market transitions made the Balinese people dependent on imported goods. In the 1960s, the Indonesian government sought to create revenue from international tourism and developed a campaign

Fig. 8.2 Four macaques cross one of the roads surrounding the Ubud monkey forest

Fig. 8.3 A macaque eats
religious offerings outside
of a café on the outskirts of
the Ubud monkey forest

to reframe Bali as "Island of the Gods" and successfully marketed its close proximity to Australia and New Zealand. These efforts were soon extended to Europeans and US Citizens creating a successful international tourism destination for decades. However, the global economic crises in 1998, and the terrorist attacks of 2002 and 2005 on popular tourist locations (nightclubs, restaurants, and shopping malls killing 222 people and injuring 400 others) had a cooling effect on tourism. After slowly rebuilding the tourism industry following the attacks, the COVID-19 pandemic has curtailed almost all travel on and off the island and has ground tourism to a virtual halt (as of mid-2021).

The fates and fortunes of the Ubud Monkey Forest are largely reliant on international tourism. The monkey forest represents a site with high degrees of human-macaque overlap which is bolstered by the revenues that are generated via entrance fees from tourists who wish to interact with the macaques. Prior to the pandemic, Ubud was consistently one of the top-rated tourist locations in Bali. The monkey forest alone attracted ~1200–3300 tourists daily (Pak Buana; former Director of the Ubud Monkey Forest; pers. comm.). Annually, tourists visited from a staggering 70+ countries. Increased tourist traffic has driven the transformation of bucolic sculpted landscapes to address the burgeoning accommodation needs.

In addition to intensifying the landscape overlap, it has resulted in close human-macaque proximity. The monkey forest had historically provided tourists with the opportunity to interact with free-ranging wild primates that includes a spectrum of engagement ranging from passive (walking among wildlife, photographing), and physical contact (interactions associated with selfies, touching, teasing, feeding) (Fig. 8.4). These interactions were moderated by the monkey forest guards, but intensified by rising monkey and tourist populations, which increased the potential

Fig. 8.4 A tourist takes a close photo of a monkey

for engagement. Recently, the monkey forest policy has prohibited physical contact between tourists and the macaques and has reduced macaque contact with guards. These modifications bring the forest closer aligned with recent International Union for the Conservation of Nature (IUCN) recommendations on responsible images of nonhuman primates (Waters et al. 2021).

8.4 Successes in the Ubud Monkey Forest

8.4.1 Education and Outreach

Managing the human dimension, public education, and communication efforts are critical to promote peaceful human-macaque interfaces. Public education programs generally improve the perceptions, attitudes, and tolerance of people toward macaques and reduce the risks associated with problematic interactions (Jones-Engel et al. 2011). One of the primary objectives defined by the Padangtegal Wenara Wana Foundation (i.e., Ubud Monkey Forest's governing council) is "Educating people about the importance of conserving the Sacred Monkey Forest's natural and cultural resources" (http://m.facebook.com/MonkeyForestUbud).

The forest management has developed a multifaceted education and marketing outreach strategy that integrates visitor information and recommendations. Each visitor is provided an informational flyer at the entrance available in 13 languages. This flyer provides information on the philosophy and history of the sanctuary, the temples, the forest, the monkeys, and goals of the organization. Throughout the forest, large visually appealing signs provide "Monkey Forest Tips" sharing safety and health information associated with human-macaque interactions. These tips (e.g., don't touch or tease the monkeys, don't feed the monkeys, don't bring or hide food or bottle, don't make eye contact, hold onto your belongings) are effectively combined with the constant recommendations made by the forest staff. These efforts are supported through the official website and social media pages that also highlight the cultural and spiritual values of this sanctuary for the Balinese people and safe practices for visitors.

8.4.2 Forest Structure Management

Deforestation and land alterations around the forest during the 1970s–1990s resulted in significant local erosion. In 1998, the first international Balinese Macaque Field school, a collaboration between Central Washington University (WA, USA), Universitas Udayana (Denpasar, Bali, Indonesia) and the temple staff, and local management team of the Ubud Monkey Forest took place. One of the key problems presented to the research team was the substantial erosion in the forest.

Between 19,982,000, the field school participants and leaders, scholars from Universitas Udayana (UNUD), and the local forest staff undertook studies and developed management plans to increase the health of the forest. The result included building retaining walls with bamboo, eventually replacing bamboo with stone walls, removing plastic debris from the soils in the forest, replanting saplings, and structurally supporting smaller or damaged trees (including removing dead trees, a long and complex cultural and religious process in addition to an ecological and practical one).

Starting in the early 2000s, the Ubud Monkey Forest management reclaimed land adjacent to the forests, purchased a parcel of land from the neighboring village of Nyuh Kuning, and reclaimed sawah fields for the purpose of replanting an adjoining forest. By 2007, these actions had resulted in a threefold expansion of the area of the monkey forest (~8 hectares in 1998 to ~21 hectares in 2018) and a marked rejuvenation of both the undergrowth and soils and the health of the larger and mid-canopy trees. Such endeavors continue through today, however, the landscape and ecological dynamics of the forest have been dramatically altered by extensive construction of tourist-related buildings inside the forest, large parking lots adjacent to it, and modifications to the pathways, open areas, and temples at the site between 2010 and the present day.

8.4.3 Reduction in Plastic Pollution

Bali faces a plastic pollution crisis. Today, plastic is prevalent throughout the numerous rivers of Bali, on Balinese beaches, and in villages and sawah fields. The transition from plant-based materials to plastic as the primary packaging for small items, combined with the positioning of the temple forest as a convenient path between villages, and the use of the temple as a center of worship resulted in a high density of plastic refuse in the forest soils. By the later 1990s, the relative number of tourists visiting the site had increased dramatically (to hundreds a day). These impacts (plastic, increased human use, erosion) resulted in damage to the soil and floral communities in the forest.

Removing access to trash is an essential strategy in the human-macaque conflict mitigation (Jones-Engel et al. 2011). Given their generalist and opportunistic diet, long-tailed macaques commonly forage for food in bins and dumps, which represents a significant source of nuisance for people and hazards for macaque health. In the last ~20 years, the Ubud Monkey Forest stakeholders and the village leadership of Padangtegal have paid close attention to waste management, especially with respect to plastic removal and recycling. By 1999, Padangtegal established a trash collection service for the community. This resulted in a reduction of the waste burning practices. The management team worked with local vendors to lower the amount of plastic pollution surrounding the site. Within the forest, macaque-proof recycling bins, daily sweeper service, and forest workers awareness about plastic-related issues helped institute a culture of plastic removal. There were three primary goals associated with this initial plastic removal: (1) increase the ecological health of the forest (plastic disrupts soil and aquatic nutrient cycling), (2) improve the health of the macaques (who occasionally ingest plastic), and (3) increase the overall beauty

Fig. 8.5 A young macaque with a bottle cap lodged in their cheek pouch

of the forest for the local Balinese and tourists by attempting to represent the environment closer to a natural state. By 2013, the forest management initiated a waste management system consisting of bins distributed throughout the site identifying organic waste, recyclable and nonrecyclable waste. The forest management built a facility to sort and process recyclables (tin and aluminum cans, glass bottles, and plastic containers) and compost organic waste (provisioned potato and banana peels, corn cobs, papaya stems, and fallen forest leaves).

Plastic bottles and bags are prohibited within the sanctuary. This is enforced by staff at forest entrances, regular forest guard warnings, and related signage. Object manipulation is very common in this population (e.g., stone handling, object play, opening bottles, unzipping backpacks of visitors) (Fuentes et al. 2011; Cenni et al. 2020) and the learning capacity of the macaques challenges forest workers to develop varying preventive measures of nuisance reduction. The mitigation of plastic bottle waste in the forest simultaneously reduces risks associated with nonfood object manipulation such as bottle caps getting stuck in cheek pouches of the macaques (Fig. 8.5). If not dislodged, the bottle caps may ultimately puncture through the cheek pouch resulting in an open wound and secondary infections.

8.5   Health and Nutrition

8.5.1   Veterinary Care

In response to the growing macaque population and tourist pressures to address the health of macaques, the sanctuary opened a veterinary clinic on the premises. Tourists visiting the site voiced concerns for macaques with open wounds, broken digits, missing tails, feet, or hands. The primary goal of the veterinary clinic is to care for and rehabilitate macaques. The clinic works with local veterinarians and faculty at the Universitas Udayana and is staffed by a small team of guards who care for sick or injured macaques during their rehabilitation. Common procedures include suturing open wounds, treating burns, removing bottlecaps from cheek pouches, addressing broken limbs, and the administration of antibiotics. Less frequently, sick and injured monkeys are sometimes transported to the clinic and anesthetized by the veterinary team.

The macaques at the Ubud Monkey Forest sustain injuries from macaque-macaque conflicts and risks associated with living in an urban environment. Macaques are treated for "naturally occurring" wounds, injuries, and pathologies. These include cuts and slashes from fighting, sickness and malaise, stiff joints, tumors, falling from trees, and seizures. It also includes injuries or wounds sustained by living in an urban environment; electrocutions, vehicular accidents (cars or motor scooters), bottle cap injuries, and infrequently, gunshot wounds.

After treatment, most macaques are returned to their social group. However, in one instance, an adult male macaque (named Nelson by the staff) lost his vision and could not safely reintegrate with his group. To ensure his safety, Nelson has lived several years in an enclosure by himself and is cared for by the veterinary staff. Although alone, Nelson has contact with macaques who approach his forest enclosure and the staff address some of his social and emotional needs by grooming him and combing his pelage. Recently, Nelson has been joined by an adult female named Tumsist who had her foot and arm amputated following an electrocution. This accident made it impossible for her to safely be re-introduced into her social group. After being sterilized she was introduced to Nelson's enclosure, and we wish them the best.

8.5.2 Nutritional Changes

One of the first studies (in 1998) conducted in collaboration with the forest management team was a detailed assessment of the specific foods being brought into the forest by tourists and being provisioned to the macaques by the management staff. This was instigated by the high levels of obesity in the macaques and the frequent and at times serious, conflicts between macaques around provisioning by the staff. The results were shocking, some adult female macaques were consuming up to 20% of their body weight in peanuts daily and the overall mix of dietary items skewed away from a balanced nutritional structure for the majority of adult macaques.

In response to the study, the collaboration resolved to reduce and restrict the amount of peanuts coming into the forest and then create a more formal and organized provisioning strategy (distribution locations, timing/amount, and feed content). The approach resulted (within ~1–2 years) in substantial decreases in obesity and severity of feeding-related conflict and improved reproductive health, with a higher birth rate and lower infant mortality for the macaques. Unfortunately, these changes combined with other health interventions across the 2000s (including the addition of a nutritionally balanced provisioning program and the opening of the veterinary clinic) resulted in the unforeseen and problematic ~450% growth rate in the macaque population across the last 20 years (~200 individuals in 2000 to ~1100 in 2020, see also Brotcorne et al. 2011).

8.5.3 Provisioning

Besides undesirable crop and garbage raiding, the deliberate food provisioning of primates by people is a key component of the interface in temple grounds, tourist sites, and urban environments. Although provisioning is generally motivated by cultural and religious intentions, when not effectively managed, it often leads to increased conflict and problematic human-macaque interactions (Orams 2002).

Among the undesirable effects, the aggressiveness of macaques toward people during provisioning is a commonly noted problem (Wheatley et al. 1996). In addition, access to provisioning areas is a known source of intra and intergroup competition for macaques (Brotcorne et al. 2015). The clumped and calorically rich nature of human foods such as bananas provided by tourists naturally increases the competition between groups (e.g., Asquith 1989). In the Ubud Monkey Forest, agonistic intergroup encounters took place regularly in the vicinity of the main provisioning areas. Globally, low-ranking groups (i.e., the groups loosing most frequently in intergroup encounters) had lower access to provisioning zones and were more peripheral compared to high-ranking groups whose home ranges were central within the monkey forest.

In an urban setting where the natural carrying capacity of the habitat is insufficient to support the nutritional needs of a large population, a delicate tradeoff has to be found between feeding sustainability to satiate the macaques and simultaneously limit the side effects of provisioning (related to social tension, macaque health, and public health risks). In the Ubud Monkey Forest, macaques are mainly provisioned by trained staff throughout the forests at feeding sites where cages are located with food (sweet potatoes, fruits, and vegetables).

Until recently, visitors were also allowed to feed the monkeys in a restricted number of locations (i.e., banana stands at the three entrances of the site). Feeding by tourists was regulated by highly trained forest guards and the vendors of the stands. These banana stands were run by villagers from Padangtegal and generated an additional source of incomes for the local community. In 2018, managers removed the banana stands and improved the provisioning system by staff workers with more diversified food items provided at a greater number of feeding areas. These changes aimed at reducing the risks associated with close physical contact during macaque-tourist feeding interactions and providing the equality of access to food between groups.

8.6   Challenges to the Ubud Monkey Forest

The tourism industry is changing the dynamics of human-wildlife interactions in rural and under-developed monkey forests across much of southeast Asia. Since tourism revenue is often re-invested to build education, transportation, and health infrastructure, local Balinese communities have a vested interest in maintaining and enriching the visitation experience for non-Balinese visitors. This is accomplished in part by promoting close human-wildlife contact. For macaques, such contact may lead to increased stress and anxiety (Maréchal et al. 2011). In addition, the density and close proximity of tourists may trigger aggression in macaques and create human-wildlife conflict.

The Ubud Monkey Forest management is tasked with balancing the inherent risk to the tourist (e.g., monkey bites and scratches) during the tourism experience without deterring tourists from returning in the future or jeopardizing the macaque population viability. This trend is further driven by local provisioning programs leading to high macaque population densities and increasing the likelihood of human-wildlife conflict.

8.6.1   Population Growth and Sterilization Campaigns

In addition to improving the nutrition and overall health of the macaques, provisioning has increased the reproductive success of the monkeys. The provisioning program has resulted in rapid macaque population growth. In 1978, the Ubud Monkey Forest consisted of one group with 31 macaques (Koyama 1981) and today there are ~1100 individuals split into eight social groups (Giraud et al. in prep). At present, the macaque density is 54 individuals/ha, and the site is overcrowded resulting in additional challenges for macaques, including growing social tension (Giraud 2015).

This population increase has led to intensified human-macaque overlap. The rapid population growth ensures that tourists will be able to see monkeys. However, the burgeoning population has led to the natural fissioning of especially large groups (including three fission events recorded over the past 5 years). It has also resulted in the expansion of group home ranges outside of the forest, rare opportunistic crop-raiding behavior in sawah fields and fruit trees, as well as expanding into neighboring hotel and restaurant premises.

Ironically, provisioning was originally implemented in part to reduce the number of macaques leaving the forest. These transformed landscapes and their use by macaques for the most part entertain tourists. However, the use of these spaces by both humans and macaques also introduced a number of challenges to local businesses including property destruction, negative human-macaque interactions (i.e., stolen possessions, and physical contact), and increased pressures on the monkey forest management to control the behavior of the macaques.

Therefore, the question of macaque population control has become important. Following in-depth discussions with the local stakeholders, a Belgium-Indonesia primatological and veterinary team (Universitas Udayana and Liège University) has begun the process of population control by enacting a multiyear sterilization campaign in order to stop the exponential growth and hold the population back at the equilibrium. Prior to the campaign, an estimation of the rate of sterilization needed has been done using a matrix population model (Caswell 2001) based on a 10-year demographics dataset (Cloutier et al. 2020). This campaign mainly focuses on multiparous females (female who have had at least one offspring), and ~ 42% (N = 136) of the reproductive females underwent endoscopic tubectomy over successive campaigns between 2017 and 2019 (Deleuze et al. 2021). This program utilizes a multidisciplinary approach combining population management with a long-term monitoring of demographics and potential behavioral effects of the interventions (Giraud et al., in prep). Following the first round of sterilizations, birth rates and population growth have started decreasing in 2020. However, the rapid demographic turnover projections of the population trends over 25 years suggest that further sterilization efforts must be maintained on a regular basis to keep the population at the equilibrium.

In addition to sterilization campaigns, land purchases by the management secure more space to accommodate macaque ranging needs. The monkey forest purchased adjacent land in the early 2000s that was used to grow *alang-alang*, a grass utilized for roofing material. They planted trees which have since grown into a forest habitat known as the "new forest" (Figs. 8.6 and 8.7). This forest extends the monkey forest southeast to the river, which the macaques cross regularly during the day. The far side of the river is characterized by very little urbanization compared to the central

Fig. 8.6 New Forest 2007. Reforested area reclaimed from sawah fields shortly after planting

Fig. 8.7 New Forest 2016. Reforested area reclaimed from sawah fields 10 years into growth

tourist hub of Padangtegal/Ubud. Instead, the ecology on that side is composed largely of rice fields and forest patches. As such, this space provides an outlet for at least two macaque groups at the monkey forest to move away from the densely populated streets, restaurants, and hotels of Ubud.

Our collaborative research at the site has perhaps been too successful. Our initial interests in 1998 were to stabilize a healthy population of macaques and improve the ecological functions of the forests situated in an urban environment. We knew then that in the future the villages of Padangtegal, Ubud, and Nyuh Kuning would become more populated thus further limiting regional resources including land and access to clean water. We also projected the rise in international tourism and the middle class throughout Asia, particularly China. However, in 1998 the Sangeh Monkey Forest was the most visited temple site and we expected this would continue. We were unprepared for Sangeh to fall out of favor with tourists and for them to turn their attention to Ubud. The ~1200–3300 daily tourists who visited prior to the pandemic overwhelm the walking paths and increase the likelihood of agonistic human-macaque interactions.

8.6.2 Aggressive Behavior and Modifications

From 1998 through the present, foreign researchers, researchers from Universitas Udayana, and the management and staff of the Ubud Monkey Forest have collaborated to develop a range of behavior modifications associated with macaque-human interactions. The years 1998–2002 were the only consecutive period where specific

168 M. E. Howells et al.

studies on human-macaque interactions were conducted, but data on the interface have been collected periodically by researchers since then. The management and staff at the Ubud Monkey Forest have been collecting bite data from all reported bites (reported by tourists) and have kept records for at least 10 years.

In 1998 and through to the present, the issue of macaques seeking food from humans, jumping on humans (primarily tourists), scratching, and biting humans has been a concern. In 1998–1999, the research collaboration and temple staff initiated a suite of intervention policies and practices to reduce such aggressive interactions. Chief among them was the active action of staff to intervene early in macaque-human conflict, and to immediately respond to macaque aggression toward humans. In the cases of highly aggressive macaques (almost always adult males), the staff would set up a series of "policing follows" of that individual. These follows consisted of one or two staff following the macaque on a daily basis and reacting with aggression (vocal, slingshot, display with sticks) toward the macaque any time they expressed aggression. Such focused efforts lasted up to multiple weeks in some cases, usually with the outcome of substantive behavior modification on the part of the macaque.

Since 2000, there has been a suggested ban on tourists feeding the macaques; however, this has been minimally enforced until recently. Instead, the temple staff have focused on behavioral intervention and control of the macaques. The overall outcome has been that the macaques are highly responsive to vocal and manual commands by the staff, but will actively attempt to circumvent staff observation/ interference when trying to obtain food from tourists. However, these interventions by staff resulted in a robust decrease in conflict and bite/scratch rates by 2002. The long-term effectiveness of these interventions is reflected in reduced aggression by macaques toward visitors. In the 12 year-period between 2002 and 2013, we observed a considerable reduction in the frequency and intensity of agonistic human-macaque interactions (Brotcorne 2014; Fuentes and Gamerl 2005). However, increasing size of the macaque population and the massive increase in tourist density through 2019 resulted in a higher overall amount of conflict as a result of further nuisances caused by macaques to local neighbors and problematic encounters between tourists and macaques. The current situation is not clear as the COVID-19 pandemic has modified tourist presence/patterns.

Human-directed aggression among the long-tailed macaques occurs not only due the large influx of humans into the monkey forest on a daily basis, but also because of the hands-on approach to wildlife-based tourism that banana selling and provisioning encourages. Until 2019, tourists were instructed to feed monkeys by allowing them to climb on their shoulder for a photo opportunity and then handing them the banana. In other cases, tourists will keep the bananas in their backpacks or pockets and try to sneak around the monkey forest with them, which results in the macaques climbing onto tourists to try to open their backpacks or pockets to get at the hidden food items. When tourists resist such behavior, the macaques may respond aggressively.

Such increased physical contact between humans and macaques, regardless of intent, increases the likelihood of an aggressive interaction. From May 2017 to

March 2018, subadult males exhibited 176 aggressive signals toward humans (Peterson, unpublished data). About 127 of these events were scored as "non-contact aggression" meaning that the macaques did not come into physical contact with the humans, and include such signals as lunges, threats, and eyebrow raises. The remaining 49 aggressive encounters involved direct physical contact including bites, scratches, and grabbing. During that same period, humans directed 17 aggressive interactions toward the macaques, demonstrating that macaque-human aggression is bidirectional.

The majority of interactions between humans and macaques do not result in contact. On the rare occasions when aggressive contact occurs, tourists can access one of the two first aid clinics that are located at the monkey forest, free of charge. The first aid clinics are staffed by nurses who treat superficial wounds and provide follow-up care instructions. For deeper wounds, the first aid clinics provide transportation to one of the local village clinics for treatment but such measures are very rare. Although tourists frequently voice concerns about contracting rabies from these encounters, there is no evidence prior to or at the time of writing that rabies has been or is present in the Ubud Monkey Forest macaques.

8.6.3 Ranging and Changing Dispersal Patterns

Prior to the COVID-19 pandemic, the Ubud Monkey Forest was a highly frequented tourist hotspot. As the sanctuary was gaining in popularity, the visitor flow has drastically increased. Whereas about 800 visitors were recorded per month in 1986, prior to the pandemic the sanctuary hosted between 1200 and 3200 visitors per pay (Pak Buana; former Director of the Ubud Monkey Forest; pers. comm.). The high tourist flow and the limited space available lead to a high macaque and tourist density which may negatively impact natural resources, macaque behavior, and their ranging patterns. For example, the macaque demographic density, the high degree of anthropogenic habitat disturbance, and the human presence appear to be key predictors of the level of social tension amongst Balinese monkeys (Brotcorne 2014). In these conditions, macaques have increasingly ranged outside of sanctuary and into tourist areas, returning regularly to the forest to feed.

The macaque home ranges have increasingly extended to forest areas not accessible to visitors. This includes climbing vertically into the forest canopy or into the riverine forest corridors that cannot be access by people. The intergroup competition for space also forces some groups to utilize less favorable human infrastructures located near the sanctuary, including hotels, homes, restaurants and shops, garden plots, and roads. As a consequence, complaints from neighbors about macaque nuisances have increased over the past years. The Ubud Monkey Forest management has attempted to manage these conflicts by developing a variety of strategies including a reforestation program, development of open buffer zones located between the forest edges and human properties, increased guard activity, and the sterilization program to control the population size in the long-term. It is worth noting that the frequency of crop raiding remained quite low in this area (0.7 crop raiding event per day in 2009). A questionnaire survey conducted in 2013 reported that 44% of farmers owning fields around the sanctuary experienced crop damage by macaques. Thirty seven percent of these farmers reported that the raiding frequency has decreased over the years and the others considered that the frequency has remained stable, demonstrating the relative efficiency of the guarding system (Brotcorne 2014).

Extended ranging patterns can also have an impact on intergroup relations when it results in intense home range overlap between groups. Previous work on two social groups with substantially overlapping home ranges found that subadult males from these groups exchanged affiliative gestures and groomed each other (Peterson et al. 2021). These findings suggest that social flexibility is central to understanding the totality of influences stemming from changes in resource or habitat use. Ecologically speaking, increases in overlapping home range use between two groups result in a greater number of individuals occupying the same space, potentially bringing them into conflict over the resources located therein.

Predictions based on conventional socioecological theory would anticipate a corresponding increase in intergroup competition for those resources, as we noted above. The observed presence of substantial intergroup affiliative behaviors in this context therefore merits explanation. Systematic daily provisioning from temple staff at the monkey forest operates as an abundant source of food that is consistently and predictably available, lowering uncertainty. Furthermore, informal provisioning of bananas from tourists occurred largely in specific locations where the banana sellers were set up, but tourists also carried them throughout the park to disperse them and make them less monopolizable. The unique food resource stability observed at the Ubud Monkey Forest may help account for the observed patterns of intergroup affiliation in subadult males (Peterson et al. 2021).

Dispersal patterns may also be influenced by this increase in range overlap. The long-tailed macaque population at the site has been observed exhibiting inconsistent dispersal patterns, atypical for the species' dispersal regime (Fuentes et al. 2011). Over a 10-month period from May 2017 to March 2018, a previous study observed three subadult males transfer from the New Forest group to the Temple group utilizing a strategy in which the three of them stayed together and supported one another in contests with resident Temple group males (Peterson, unpublished data). This transfer occurred over a three-month subset of the entire sample period and included affiliations with some resident males and aggression directed at others. During this same study period, a single subadult male transferred from the Temple group to the New Forest group. This individual did not have peer support and engaged in a mixture of affiliative and submissive behaviors with resident subadult males in New Forest group. Each of these dispersal events were still underway in March 2018 when data collection for this project ended. The extended timeframe for these dispersals, and the variable patterns through which they occurred, speak to the importance of understanding dispersal as a process not an outcome (Strum 2012), and why the ecological and demographic conditions at the Ubud Monkey Forest are conducive to such inconsistent and variable dispersal events.

One method to mitigate the social and psychological stress experienced by macaques would be to further expand the available space for the growing population. This could include planting more trees to provide an arboreal refuge, and purchasing more land. Both of these solutions have been employed and the coordinated programs to remove plastic pollution, reduce erosion, and plant a diversity of trees have been successful. However, we recognize that land is extremely expensive and difficult to obtain in Bali especially in top tourist destinations.

8.7   Research and Teaching

Ongoing collaborations between the Ubud Monkey Forest and the Universitas Udayana in Bali, the University of Liège (Belgium), the University of Lethbridge (Canada), East Carolina University (USA), Notre Dame University (USA), Princeton University (USA), and the University of North Carolina-Wilmington (USA) have transformed the site into an epicenter of primatological and ethnoprimatological research.

8.7.1   Balinese Macaque Ethnoprimatology Field School

This site has been the long-term home of a mixed methods ethnoprimatological field schools as well as the site of multiple theses and dissertations. In 1998, Agustín Fuentes began the Balinese Macaque Field School out of Central Washington University. This field school aimed to expose undergraduate and graduate students from a variety of backgrounds to the methods and theory associated with ethnoprimatology. This program developed into a collaboration (1999–2003) with the University of Guam's Professors Rebecca Stephenson and Hiro Kurashina. The University of Guam team continued to teach the field school from 2003 to 2008. Two of the previous attendees, James E. Loudon and Michaela E. Howells took over the field school in 2016 hosting the program through East Carolina University and University of North Carolina-Wilmington.

This program provides experiential learning opportunities within the regional south of the United States and creates an opportunity for these students to contribute in a meaningful way to the community that they are working within (Howells et al., 2022). Students engaged in this program collect data for forest management as a service component. This includes demographics of tourists, tourist behavior, and monkey censusing. In addition, students work directly with local stakeholders to identify ecological challenges facing the forest and community and develop quantitative and qualitative projects to address these obstacles. Multiple attendees of this field school continued their education and pursued careers in related fields. Attendees continued their training in fields as diverse as anthropology, biology, genetics, global health, wildlife ecology, law enforcement, education, and social justice.

This program remains one of the few ethnoprimatology field schools in the United States. It is also consistently one the most diverse biological anthropology field schools due to active recruitment of groups under-represented in anthropology. For instance, from 2016 to 2019, 23.5% of the attendees self-identified as ethnic minorities (including African Americans, Native Americans, Hispanics, Latinx, Pacific Islanders, and Southeast Asians), and 76.5% of the attendees were women.

8.7.2 Additional Scientific Training for Developing STEM Professionals

In addition to the USA and European-based scholars who conducted their thesis/ dissertation work (including co-authors Fany Brotcorne and Jeffrey V. Peterson), multiple Balinese students were trained and received advanced university degrees in association with the field school. These most notably included Universitas Udayana professors, co-authors Dr. I Gusti Arta Putra (Faculty of Animal Husbandry) and Dr. I Nengah Wandia (Director of the Primate Division of Natural Resources and Environment Research Center).

8.7.3 Collaborative Research and Outcomes

Over the course of more than 30 years, primatologists, anthropologists, and other scholars from the USA (including Guam), Canada, multiple European countries, and Japan have engaged in research collaborations with the Ubud Monkey Forest in collaboration with researchers at Universitas Udayaya in Denpasar, Bali. Under the permissions granted by the Indonesian Federal, regional, and local governments, collective research agreements have resulted in collaborative endeavors that lasted from weeks to years since the early 1990s (see Wheatley 1999; Fuentes et al. 2011). The heart of such collaborations are the relationships built between the foreign researchers and the Balinese in charge of and working in the monkey forest and the scholars from Universitas Udayana who are the main interlocutors and actors in the year-round on the ground assessment of, and interventions into, the health of the macaques at the site.

For instance, the primatology team from University of Liège (Belgium) and the Primate Division of Natural Resources and Environment Research Center from Universitas Udayana have been conducted successive collaborative research projects since 2009 on Balinese macaques. This long-term collaboration is based on the agreement of co-authored publications, a Memorandum of Understanding between both institutions and Letters of Intention for specific research projects such as the sterilization program in Ubud. In addition to a common interest in promoting the mutual cooperation for academic and research purposes, the educational scope of this collaboration involves teaching activities such as lectures, workshops, and training to allow knowledge and technology transfer between research teams (e.g., a multiyear training of the endoscopic tubectomy surgery techniques involving the Belgian and Indonesian veterinary teams).

The majority of foreign research at the site has been part of ongoing collaborative projects. Much of that work (particularly the most successful projects) have been premised on the needs of the monkey forest management in regard to the assessment, control and health of the macaques, and the forest.

This emphasis on the local community's and forest management's needs and interests in foreign researcher's project designs has resulted in substantial ecological and structural changes at the forest site and benefited the health of the macaques and the forest. The long-term research outcomes have also facilitated increases in the economic benefits for the management and the local community. However, some outcomes from research and the resultant recommendations by foreign and Balinese researchers have also had unforeseen and at times problematic results.

8.7.4 Current and Future Research at the Site

Research projects in Ubud Monkey Forest encompass behavioral ecology of urban macaques (i.e., anthropogenic impacts on behavior, ecology, and demographics), management strategies of the human-macaque conflict including an assessment of their effectiveness, and the potential side effects of human interventions, ethnoprimatology, cultural primatology, and veterinary medicine research.

This macaque population has sparked particular attention for researchers over the years. Initially designated as a food-enhanced population of long-tailed macaques, their behavior was thought to differ from other populations of this species because so much of their dietary intake was provided by humans and as a result, they experienced different sets of ecological pressures (Wheatley et al. 1996). The Ubud macaques also became somewhat of a symbolic population exemplifying the religious reverence Balinese people held for long-tailed macaques (Wheatley 1999), and how that strong cultural connection has a foundation in a long history of shared ecological spaces (Fuentes et al. 2005, 2011; Loudon et al. 2006). Over the years, tourism to Ubud Monkey Forest increased and primatologists became interested in understanding the dynamics of close and frequent contact between humans and the macaques from an ecotourism management perspective (Fuentes et al. 2007) as well as with respect to the foundational principles of Balinese Hinduism overseeing the relationship between local Balinese people and the macaques as the latter become increasingly popular tourism objects (Fuentes 2010).

More recently, researchers studying the behavioral ecology of the Ubud Monkey Forest macaques in a comparative framework found that their behavior differed in important ways from other food-enhanced long-tailed macaque populations in Bali. For instance, the Ubud Monkey Forest population was found to have larger group sizes and higher ratios of adult females to adult males than the Uluwatu and Taman Nasional Bali Barat populations (Brotcorne 2014). Additionally, subadult male

macaques at the Ubud Monkey Forest spent less time grooming and resting in proximity to each other, and exchanged fewer affiliative gestures, than those at Uluwatu
(Peterson et al. 2021). These recent studies suggest that unique social and demographic factors beyond food enhancement are important for contributing to the
unique behavioral profile exhibited by the long-tailed macaque population at the
Ubud Monkey Forest.

Finally, the population of long-tailed macaques is the only population in Bali to
exhibit stone handling, a socially transmitted and cultural behavior that involves
object play with stones following specific patterns and occurring in specific contexts
(Pelletier et al. 2017; Cenni et al. 2020). For these reasons, the Ubud Monkey Forest
has been, and continue to be, of high interest to primatologists investigating a range
of complex demographic, ecological, and behavioral patterns.

There are a number of future research projects planned at the site. These include
assessing the impacts of humans on the nutritional ecology of the macaques, an
assessment of the nutritional properties of the provisioned and natural foods consumed by the macaques, the role of social media on tourist interactions with monkeys, and the stable isotope ecology of the forest ecosystem. As a continuation of
the early work on human-macaque interconnections and zoonotic transmission, we
also aim to investigate the gut microbiomes of macaques and humans at the site with
an eye toward understanding the transmission of commensal microbial communities between the two primate species.

8.8 Looking to the Future

While human population growth and urbanization continue to expand, we face an
increasing need for multidisciplinary studies at the human-macaque interface as a
combined effort to assess the most effective, adapted, and ethical management strategies of conflict mitigation and co-existence (Priston and McLennan 2013; Waters
et al. 2021). Macaques in the Ubud Monkey Forest illustrate this growing phenomenon whereby NHPs and humans are co-participants in a shared ecosystem (Fuentes
2010), may result in conflict when NHPs proliferate in anthropogenic environments.
Fertility control is increasingly used to keep in check uncontrolled NHP population
growth as it represents an ethical alternative to culling and translocation (e.g., Shek
and Cheng 2010). However, such programs require a holistic and long-term monitoring approach, as currently conducted at this site, to ensure management efficiency and population viability, while understanding the implications for primates
(Giraud et al., Effect of infant presence on social networks of sterilized and intact
wild female Balinese macaques (*Macaca fascicularis*), in prep).

It is worth noting that a solid foundation of knowledge of the population and the
causes underpinning the human-macaque conflict is a cornerstone of the decision
process prior to any program implementation (Jones-Engel et al. 2011; Brotcorne
et al. 2018). Finally, it is also important to emphasize that conflictual issues are
naturally multifaceted. Therefore, fertility control, when relevant at a site, should be

considered as a part of an integrated management strategy involving natural habitat restoration and environment management, food provisioning control, and educational programs available to local stakeholders.

Forthcoming work at the Ubud Monkey Forest will require plans to accommodate for future pandemics. The global outbreak of the coronavirus has hurt the tourism industry in Bali and required the monkey forest to institute policies to ensure the health of their staff and reduce the likelihood of the transmission of the disease. In 2020, the management closed the forest to the public from March to November, resulting in the loss of revenue generated from entrance fees paid by tourists. Despite these measures, the staff were fully paid, and the macaques continued to be provisioned.

The COVID-19 outbreak is a reminder of the possibility of future pandemics. As humans and wildlife continue to live increasingly in close association, the possibility of zoonotic transmissions that could result in pandemics continues to rise (Lappan et al. 2020). Global pandemics which reduce or eliminate tourist revenues can be especially problematic for populations of urban monkeys in Southeast Asia and beyond. Many of these monkey populations are large and expansive and their health and social structures largely depend on a thriving tourist industry which interjects monies into local economies and lowers the pressure on the people who they live among.

One very thin silver lining of the COVID-19 pandemic has illustrated the flexibility that the monkey forest management has utilized to navigate the complications with the disease and its associated economic hardships. These solutions should be re-enforced, and a series of contingency plans should be developed and be readily employed on the event that another pandemic may emerge. This approach should include the expertise of the Ubud Monkey Forest management and staff, local stakeholders, the governing institutions of Bali and Indonesia, and the faculty at the Universitas Udayana. Western researchers may act as advisers if their assistance is useful and invited.

8.9 Conclusions

The Ubud Monkey Forest is characterized by a series of successes and pitfalls. We are still learning. Nonetheless, the robust collaborations with Balinese, Canadian, European, and US universities, Indonesian and regional Balinese governments, and local people have contributed to our understandings of primate tourism, ethnoprimatology, and human-NHP interconnections. The Ubud Monkey Forest acts as an excellent example of how to use tourist revenues to stabilize and improve the health of macaques, conserve forest fragments that are considered sacred or important to local people and primates, and incorporate monies into local economies to improve the lives and livelihoods of people who share their landscapes with their NHP kin. We assert that this approach could benefit other monkey forests across Bali and throughout Southeast Asia. However, to do so, this requires support by the local people, cultural competency, constant lines of communication between all parties, an appreciation of historical contexts that may be specific to that region, and a healthy understanding of religious and cultural institutions of the local people.

References

Asquith PJ (1989) Provisioning and the study of free-ranging primates: history, effects, and prospects. Am J Phys Anthropol 32:129–158. https://doi.org/10.1002/ajpa.1330320507

Bali Tourism Office (2017) Bali Province central statistics agency. Republic of Indonesia, Denpasar

Bell C (2019) The new age tourism band-wagon in Ubud, Bali: Eat, Pray Love! South Asian Res J Arts Lang Lit 1(3):74–82

Brotcorne F (2014) Behavioral ecology of commensal long-tailed macaque (*Macaca fascicularis*) populations in Bali, Indonesia: impact of anthropic factors (PhD. Dissertation). University of Liège, Liège, Belgium

Brotcorne F, Wandia IN, Rompis ALT, Soma IG, Suatha IK, Huynen MC (2011) Recent demographic and behavioral data of *Macaca fascicularis* at Padangtegal, Bali, Indonesia. In: Gumert MD, Fuentes A, Jones-Engel L (eds) Monkeys on the edge. Ecology and management of long-tailed macaques and their interface with humans. Cambridge University Press, Cambridge, pp 180–182

Brotcorne F, Fuentes A, Wandia IN, Beudels-Jamar RC, Huynen MC (2015) Changes in activity patterns and intergroup relationships following a significant mortality event in commensal long-tailed macaques (*Macaca fascicularis*) in Bali, Indonesia. Int J Primatol 36:548–566

Brotcorne F, Broens D, Delooz S, Giraud G, Wandia IN, Huynen MC, Poncin P (2018) Analyser les avantages et inconvénients des stérilisations de primates en milieux anthropisés: Une étude de cas des macaques balinais. Rev Primatol 9:20

Caswell H (2001) Matrix population models. Construction, analysis and interpretation. Sinauer Associates, Sunderland, MA

Cenni C, Casarrubea M, Gunst N, Vasey PL, Pellis SM, Wandia IN, Leca J-B (2020) Inferring functional patterns of tool use behavior from the temporal structure of object play sequences in a non-human primate species. Physiol Behav 222:112938

Cloutier F, Schtickzelle N, Giraud G, Wandia IN, Brotcorne F (2020) Calibrating a sterilization program using demographic modelling: the case of Balinese macaques (*Macaca fascicularis*). Paper accepted at the International Society of Primatologists Congress, Quito

Deleuze S, Brotcorne F, Polet R, Soma G, Rigaux G, Giraud G, Cloutier F, Poncin P, Wandia IN, Huynen MC (2021). Tubectomy ofPregnant and Non-pregnant Female Balinese Macaques (*Macaca fascicularis*) with postoperative monitoring. Frontiers in veterinary science, 1008

Dore KM, Riley EP, Fuentes A (2017) Ethnoprimatology: a practical guide to research at the human-nonhuman primate interface. Cambridge University Press, New York

Fuentes A (2010) Natural cultural encounters in Bali: monkeys, temples, tourists, and ethnoprimatology. Cult Anthropol 25(4):600–624

Fuentes A (2012) Ethnoprimatology and the anthropology of the human-primate interface. Annu Rev Anthropol 41:101–117

Fuentes A (2013) Social minds and social selves: redefining the human-alloprimate interface. In: Corbey R, Lanjouw A (eds) The politics of species: reshaping our relationships with other animals. Cambridge University Press, Cambridge, pp 179–188

Fuentes A, Gamerl S (2005) Disproportionate participation by age/sex classes in aggressive interactions between long-tailed macaques (*Macaca fascicularis*) and human tourists at Padangtegal monkey forest, Bali, Indonesia. Am J Primatol 66(2):197–204

Fuentes A, Southern M, Suaryana KG (2005) Monkey forests and human landscapes: is extensive sympatry sustainable for *Homo sapiens* and *Macaca fascicularis* on Bali. In: Patterson JD,

Wallis J (eds) Commensalism and conflict: the primate-human Interface. American Society of Primatologists, Norman, pp 168–195

Fuentes A, Shaw E, Cortes J (2007) A qualitative assessment of macaque tourist sites in Padangtegal, Bali, Indonesia, and the Upper Rock Nature Reserve, Gibraltar. Int J Primatol 28:1143–1158

Fuentes A, Rompis AL, Putra I, Watiniasih NL, Suartha IN, Soma IG, Wandia IN, Putra IDKH, Stephenson R, Selamet W (2011) Macaque behavior at the human-monkey interface: the activity and demography of semi-free-ranging *Macaca fascicularis* at Padangtegal, Bali, Indonesia. In: Gumert MD, Fuentes A, Jones-Engel L (eds) Monkeys on the edge. Ecology and management of long-tailed macaques and their interface with humans. Cambridge University Press, Cambridge, pp 159–179

Giraud G (2015) Relation between social tension and demographic density of commensal long-tailed macaques (*Macaca fascicularis*) in Bali (Indonesia). Master thesis, University of Liège, Belgium

Giraud G, Sosa S, Huynen MC, Deleuze S, Wandia IN, Hambuckers A, Poncin P, Brotcorne F (in prep) Effect of infant presence on social networks of sterilized and intact wild female Balinese macaques (*Macaca fascicularis*)

Gumert MD, Fuentes A, Engel G, Jones-Engel L (2011) Future directions for research and conservation of long-tailed macaque populations. In: Gumert MD, Fuentes A, Jones-Engel L (eds) Monkeys on the edge. Ecology and management of long-tailed macaques and their interface with humans. Cambridge University Press, Cambridge, pp 328–353

Howells ME, Woolard KL, April TB, Bender R, Loudon JE (2022) Is there a difference in student physical activity between afield school and a traditional classroom setting? American Journal of Human Biology. https://doi.org/10.1002/ajhb.23799

Jones-Engel L, Engel G, Gumert MD, Fuentes A (2011) Developing sustainable human-macaque communities. In: Gumert MD, Fuentes A, Jones-Engel L (eds) Monkeys on the edge. Ecology and management of long-tailed macaques and their interface with humans. Cambridge University Press, Cambridge, pp 295–326

Koyama N (1981) Socio-ecological study of crab-eating monkeys in Indonesia. Kyoto Univ Overseas Rep Stud Indonesian Macaque 1:1–10

Lansing JS, Thurner S, Chung NN, Coudurier-Curveur A, Karakas C, Fesenmyer KA, Chew LY (2017) Adaptive self-organization of Bali's ancient rice terraces. Proc Natl Acad Sci USA 114(25):6504–6509

Lappan S, Malaivijitnond S, Radhakrishna S, Riley EP, Ruppert N (2020) The human–primate interface in the new normal: challenges and opportunities for primatologists in the COVID-19 era and beyond. Am J Primatol 82(8):e23176

Loudon JE, Howells ME, Fuentes A (2006) The importance of integrative anthropology: a preliminary investigation employing primatological and cultural anthropological data collection methods in assessing human-monkey co-existence in Bali, Indonesia. Ecol Environ Anthropol 2(1):2–13

Maréchal L, Semple S, Majolo B, Qarro M, Heistermann M, MacLarnon A (2011) Impacts of tourism on anxiety and physiological stress levels in wild male Barbary macaques. Biological Conservation 144(9):2188–2193

Orams MB (2002) Feeding wildlife as a tourism attraction: a review of issues and impacts. Tour Manag 23:281–293. https://doi.org/10.1016/s0261-5177(01)00080-2

Pelletier AN, Kaufmann T, Mohak S, Milan R, Nahallage CAD, Huffman MA, Gunst N, Rompis A, Wandia IN, Putra IGAA, Pellis SM, Leca J-B (2017) Behavior systems approach to object play: stone handling repertoire as a measure of propensity for complex foraging and percussive tool use in the genus *Macaca*. Anim Behav Cogn 4:455–473

Peterson JV, Fuentes A, Wandia IN (2021) Male-male affiliation varies between populations in subadult long-tailed macaques. Acta Ethol 24:9–21

Priston NEC, McLennan MR (2013) Managing humans, managing macaques: human-macaque conflict in Asia and Africa. In: Radhakrishna S, Huffman MA, Sinha A (eds) The macaque connection. Cooperation and conflict between humans and macaques. Springer, New York, pp 225–251

Radhakrishna S, Huffman MA, Sinha A (2013) The macaque connection. Cooperation and conflict between humans and macaques. Springer, New York

Riley EP (2019) The promise of contemporary primatology. Routledge, New York

Shek CT, Cheng WW (2010) Population survey and contraceptive/neutering programme of macaques in Hong Kong. Honk Kong Biodivers 19:4–7

Sueur C, Petit O, De Marco A, Jacobs AT, Watanabe K, Thierry B (2011) A comparative network analysis of social style in macaques. Anim Behav 82(4):845–852

Vickers A (2019) Creating heritage in Ubud, Bali. Wacana 20(2):250–265

Waters S, Setchell JM, Maréchal L, Oram F, Wallis J, Cheyne SM (2021) Best practice guidelines for responsible images of non-human primates. A Publication of The IUCN Primate Specialist Group Section for Human-Primate Interactions. Retrieved from https://humanprimateinteractions.wpcomstaging.com/wp-content/uploads/2021/01/HPI-Imagery-Guidelines.pdf

Wheatley BP (1999) Sacred monkeys of Bali. Waveland Press, Prospect Heights

Wheatley BP, Putra DKH, Gonder MK (1996) A comparison of wild and food-enhanced long-tailed macaques (*Macaca fascicularis*). In: Fa JE, Lindburg DG (eds) Evolution and ecology of macaque societies. Cambridge University Press, Cambridge, pp 182–206

Wijana N, Wesnawa IGE (2018) The mapping of rare plant species distribution in monkey forest, Ubud, Gianyar, Bali. Komunikasi Geografi 19(1):23–30

Chapter 9
Primate Tourism on Java: 40 Years of Ebony Langur Viewing in Pangandaran from Homestay Visits to Mass Tourism

Vincent Nijman

Abstract I provide a narrative of primate (eco-)tourism over a 40-year period at the Pangandaran peninsula, highlighting research on ebony langurs *Trachypithecus auratus*, tourism and its interaction. When the first studies on langurs were conducted from 1970 to 1990s, tourism in Pangandaran could be described as "eco-", although the term was not widely in use then. Around 500,000 foreign and domestic tourists visited the isthmus, stayed with local people or in family-run hotels and ate locally prepared food. In 2006, Pangandaran was hit by a tsunami and the wooden and bamboo cafes, shops and homestays were destroyed. Figuratively, the tsunami also washed away the eco in Pangandaran's tourism. A major rebuild took place, focussing on mass tourism from within Java. Now the area is dominated by high-rise hotels catering for large groups and organised tours; in 2018, there were 4.1 million visitors, 99.8% from within Indonesia. Over these four decades, a population of ~150 ebony langurs live in the most visited parts of the Nature Tourism Park and the Strict Nature Reserve where they come into daily contact with thousands of tourists. Roads, picnic areas, parking lots and road-side tree lines are part of the langur's home range, and these are shared day in day out with people. While the number of tourists has dramatically increased, the population of langurs has remained more or less stable over time. Other than generating financial benefits for local people, the current tourism activities in Pangandaran do not follow the basic tenets of ecotourism, and at best a walk into the forest can be described as a short ecotourism experience as part of a multi-purpose trip. It appears that their arboreal nature buffered the langurs from the most obvious negative effects of mass tourism, but tourism in Pangandaran has had a negative impact on the environment for both people and wildlife.

Keywords Indonesia · Conservation · Wildlife · Lutung · Ecotourism

V. Nijman (✉)
Anthropology and Geography, Oxford Brookes University, Oxford, UK
e-mail: vnijman@brookes.ac.uk

179

9.1 Introduction

I am as you will see now commencing my retreat westwards. I have left the wild and savage Moluccas & New Guinea for Java the garden of the East & probably without any exception the finest island in the world.

  Good roads regular posting stages & regular inns & lodging houses all over the interior' make for a happy naturalist.

  Alfred Russel Wallace (WCP375: letter home to his mother, 20 July 1861)

Java is one of the premier tourism destinations in Indonesia, and indeed Southeast Asia. The island's largest city is the modern, sprawling Jakarta, the nation's capital, where one can find the National Museum, the Old Town (*Kota Tua*) with Dutch colonial buildings, as well as five-star hotels and massive shopping malls. While Java is presently home to one of the largest Muslim concentrations in terms of population (after Pakistan, India and Bangladesh), the ninth century temples of Borobudur (the world's largest Buddhist temple) and Prambanan or Rara Jonggrang (an expansive Hindu complex) are testament of a different past. The botanical gardens of Bogor (*Kebun Raya*) and the sea of sand (*lautan pasir*) high near the top of Mt. Bromo Tengger are some of the easily accessible destinations for nature lovers. In terms of tourism, what is true now was also true 160 years ago, as reflected in the quote from the famous naturalist Alfred Russel Wallace from a letter to his mother after he had spent several years in the eastern Indonesia.

The International Ecotourism Society defines ecotourism as "responsible travel to natural areas that conserves the environment, sustains the well-being of the local people, and involves interpretation and education". Principles of ecotourism are about uniting biodiversity conservation, communities, and sustainable travel; and its implementation, participation, and marketisation should build environmental awareness and respect, generate financial benefits for local people, involve and operate low-impact facilities, and provide direct financial benefits for conservation. One alternative to ecotourism is mass tourism, where tourists travel in large groups on pre-scheduled tours in an organised manner; this is generally perceived as less environmental friendly. However, as explained by Weaver (2001), ecotourism also can include travellers that embark on short ecotourism experiences as one component of a multi-purpose trip, and as such, ecotourism and mass tourism are not always contradictory. Ecotourism travellers that are also part of the mass tourism industry are usually associated with a "steady state sustainability" or leaving the area (and presumably the species that live in it) in the same condition as when they arrived, rather than with traditional ecotourists that want to leave the places they visit in a better or improved state when they leave.

While a lot of the discussion on mass tourism and ecotourism focusses on international tourists, i.e., tourists that spent anywhere between 24 h and 6 months in a country that is not their own, it is worth noting that domestic tourism is on the rise. In many countries, including Indonesia, the number of domestic tourists greatly exceeds the number of international tourists (see also Ghimire 2013). In its efforts to promote tourism, the government of Indonesia is not only set to increase the number of foreign visitors but is also encouraging more domestic tourism and, as stated by Gunawan (1996), for Indonesians to "become tourists in their own country". This domestic tourism is not evenly distributed. Winastuti (2020) suggests that in the year 2017, there were 270 million domestic tourists (an individual can be a tourist more than once a year), and the province of West Java—one of 34 provinces—received the one sixth of the domestic tourists. Only a small number of these tourists can be considered ecotourists. Discussion on ecotourism in Indonesia often focus on a limited number of specific sites, most of them national parks, including Gunung Leuser National Park in Sumatra (Cochrane 1996; Siburian 2006), Komodo National Park in the Lesser Sunda Islands (Cochrane 1996), Halimun-Salak National Park on Java (Dalem 2002; Nakashima 2001), and Bunaken and Tangkoko DuaSaudara National Parks in Sulawesi (Ross and Wall 1999; Pangemanan et al. 2012; Kinnaird and O'Brien 1996). Heavily biased to islands other than Java, these are not the parts of Indonesia that most of the domestic tourists visit (Winastuti 2020).

As Indonesia's political, economic and industrial centre, Java is one of the most densely populated areas in the world, with some 140 million inhabitants living at an average density of over 1000 people km^{-2}. Forest has been replaced by a mosaic of cities and villages, agricultural land, cash-crop plantations such as coffee *Coffea* sp. and tea *Camellia sinensis*, and forest plantations including teak *Tectona grandis*, Sumatran pine *Pinus merkusii* and rubber *Hevea brasiliensis*. Natural forests are mostly found in isolated patches in the mountains, as well as a small number of isolated lowland areas. Deforestation has a long history on Java, and by the end of the nineteenth century, the natural forest was severely fragmented, showing a pattern very similar to that seen today (Whitten et al. 1996). The distribution pattern of primates on Java is determined by the severe degree of forest fragmentation as well as by the climate (primarily rainfall, having its effect on forest type). The extant primate community of Java comprises five species (Nijman 2013). Three are endemic to the island, i.e., Javan gibbon *Hylobates moloch* and the grizzled langur *Presbytis comata* both found in the western part of the island, and the Javan slow loris *Nycticebus javanicus* that is distributed over all parts of the island. The ebony langur *Trachypithecus auratus*, besides occurring on Java, also occurs on the smaller islands of Bali and Lombok to the east (Nijman 2000). The fifth species, the long-tailed macaque *Macaca fasicularis* has a wider distribution including much of the Southeast Asia. In the past species such as pig-tailed macaques *M. nemestrina*, siamang *Symphalangus syndactylus* and orangutan *Pongo* spp. were present on the island as well (Whitten et al. 1996). Of the diurnal primates that occur on Java both the long-tailed macaques and the ebony langurs are the ones that are easiest seen by tourists. Not only is their geographic range much larger than the gibbon and the grizzled langur, but they are less confined to rainforest and occur in more open and fragmented habitats, often close to human habituation.

I here give an overview of tourist development and research on ebony langurs and tourism in Pangandaran on the south coast of Java. Over a 40-year period, it changed from being characterised by small hotels and hostels catering to individual tourists and smaller parties to it now being dominated by high-rise three- and four-star hotels catering for large groups and organised tours. Pangandaran is one of the first areas where Javan primates were studied in the late 1970s, and albeit intermittently with lengthy gaps and involving different research teams, these studies have continued to the present day. Pangandaran also is one of the several well-known tourist sites in Java. For a long time, it was part of the main itinerary of foreign tourists and more recently it is heavily promoted as a suitable alternative to Bali for domestic tourists. Twenty-five years ago, Whitten et al. (1996) already commented that "This reserve and adjacent tourist park would be rather insignificant were it not for its position at the end of a beach-lined isthmus fringed with hotels and homestays. The park in fact receives more visitors than any other conservation area in Indonesia." As I will show, Pangandaran has seen some dramatic changes over the last 40 years, with many more tourists visiting now than when Whitten et al. (1996) wrote their overview and with many infrastructural developments impacting the area. I will provide a narrative where this increase in tourism intersects with the ecology, management and ultimately conservation and welfare of the langurs.

9.2   Methods

9.2.1   Study Area

When Pangandaran was mentioned before 2012, this almost invariably referred to the isthmus, the village of Pangandaran (possibly also the neighbouring villages of Pananjung and Babakan), the Nature Tourism Park and/or the Strict Nature Reserve. Pangandaran was part of Ciamis regency (*kabupaten*) but in 2012 this regency was split into two, with the northern part retaining the name Ciamis and the southern part was given the name Pangandaran (with Parigi as its new capital). The Pangandaran regency is 1680 km² large and the Pangandaran peninsula only makes up a very small proportion of this. Whereas in the past when reference was made to ebony langurs or tourism and Pangandaran, it always referred to the peninsula (other parts were referred as Ciamis) but now it can also refer to other sites within the regency. I here refer to Pangandaran in the narrow sense.

Pangandaran is geographically divided into three sections. First, there is the isthmus where the hostels, hotels, restaurants and houses are; this is where the majority of people spent their days (Fig. 9.1). Second and third, there is the peninsula that is largely covered in forest, which is subdivided into the Nature Tourism Park (*taman wisata alam*) and the Strict Nature Reserve (*cagar alam*). The Nature Tourism Park and the Strict Nature Reserve are mostly forested, but there are open spaces (e.g., the grazing grounds in the Strict Nature Reserve or the picnic areas and area surrounding the administrative buildings (*kantor*) in the Nature Tourism Park), whereas some of the forest extents outside the Strict Nature Reserve and Nature Tourism Park. The latter takes the form of large, isolated trees that are still within a short distance from the forest edge—for instance, in and around one of the car parking areas—or in the form of lines of trees (mainly sea almond *Terminalia catappa*)

Fig. 9.1 Pangandaran, tourism and langurs. From top, clockwise. Overview of Pangandaran showing the isthmus with the village leading to the Nature Tourism Park and the Strict Nature Reserve (2014); a 4-floor hotel that opened in 2016, with the right background trees in the Nature Tourism Park (2016), ebony langurs *Trachypithecus auratus* on the ground in the Nature Tourism Park (2011), tourists on the beach, with the background lines of sea almond trees *Terminalia cata-ppa* that are used by ebony langurs (2018); the eastern beach looking south towards the Nature Tourism Park (1973), ebony langurs in the Strict Nature Reserve (2011). Photos all licenced under CC BY-NC-SA 2.0

along the western and eastern beaches (Fig. 9.1). Geographically, I here focus on the southernmost part of the isthmus, where the Nature Tourism Park, the Strict Nature Reserve and the build-up area meet. Here the interaction between wildlife and people and economic development and nature protection is the strongest; several groups of langurs, long-tailed macaques and Javan deer *Rusa timorensis* are found in this area. In terms of the langurs, I focus on what was labelled as Troops C and A in the 1970s (Brotoisworo 1983), Group 3 and Group 31 in the 1980s and 1990s (e.g., Kool 1989, 1992, 1993) and group K (Kantor) and group A in the 2010s (Tsuji et al. 2013, 2015, 2016, 2017). It was not always possible to determine to what group the langurs under observation belonged. In addition, other groups were present.

One of the more significant events that shaped Pangandaran and one that had a massive influence on the nature and type of tourism was the tsunami that hit the area on 17 July 2006 (Reese et al. 2007). The up to 5-meter-high waves flooded Pangandaran up to 400 m inland. The wooden or bamboo cafes, shops and home-stays along the waterfront and up to 20 or 30 m inland were washed away, and there was severe damage to almost all structures within several hundred meters of the waterfront. The tsunami resulted in a major rebuild of Pangandaran with a focus on mass tourism. Pre-tsunami Pangandaran was characterised by small hotels and hostels catering to individual tourists and smaller parties, whereas now it is dominated by high-rise three- and four-star hotels catering for large groups and organised tours. The new hotels were initially erected predominantly on the western beachfront, but now they are a feature throughout the isthmus (Nijman 2021).

9.2.2 Data Collection

Over the last 25 years, I have made 12 visits to Pangandaran (1995, twice in 1997, 1999, 2004, twice in 2012, 2013, 2015, 2016, 2018, 2019). Each visit lasted between 2 and 4 days, similar to that of many tourists to Pangandaran, totalling 34 days. During the visits in the 1990s and early 2000s, the mornings were spent primarily collecting data on the langurs in the Nature Tourism Park, and I would spend the remainder of the time on the isthmus. During the later visits, most time was spent on the isthmus, but I ensured that each time I did visit the peninsula and the Nature Tourism Park at least once. On most days, the langurs were observed on and from the southern tip of the isthmus. In addition to Pangandaran, I have studied ebony langurs for extensive periods at other sites in Java, Bali and Lombok so I am familiar with their behaviour and ecology (e.g., Nijman 2000, 2015a, b, 2019). For data on the langurs, I relied largely on studies conducted by others; almost all of these studies are very much centred on the langurs themselves, and little attention is paid to the interaction between them and the people (or if it is rarely made specific).

To put the studies of the langurs and tourism in context of other research conducted in and around Pangandaran, I searched for published papers and reports from the area. Following the Preferred Reporting Items for Systematic review and Meta Analysis (PRISMA) statement and procedures outlined in Moher et al. (2009), in December 2020 I conducted a systematic search to identify relevant publications on Google Scholar. I used "Pangandaran" focussing on review and research articles published between 1970 and 2020. I then scanned all article titles and abstracts (when available) and applied exclusion criteria to remove articles limited in scope to the following: the topics had to deal with (a) tourism, including domestic tourism and tourism potential, (b) physical geography, the physical landscape and the tsunami, (c) botany and vegetation science, including studies of the vegetation used by certain wildlife; agricultural studies were excluded, (d) primates and primatology, (e) studies on terrestrial animals other than primates and (f) fisheries. When a particular paper could be included on two or more categories, I selected the one that was most fitting. For each of five decades (1970–1971, 1980–1989, etc.), I tallied the number of publications and ranked topics by the number of times the papers were cited. In recent years, many more Indonesian language scientific publications have become available online, but since these were not available for earlier periods. I restricted this part of the search to English language papers only.

In December 2020, I searched for photos on Google using the keywords Pangandaran in combination with *monyet*, *kera*, macaque (this should return mainly photographs of long-tailed macaques), *lutung*, langur (this should return mainly langurs) and monkey (this can return both species). I tallied the number of photographs of the two species that showed up and, arbitrarily, stopped tallying when I decided too many irrelevant photographs were returned.

All prices quoted here were corrected for inflation using an online inflation calculator (www.oanda.com) to December 2020 (so that Rp1,000 in 1992 equalled Rp10,347 in 2020), and then converted to US$ using a value of Rp14,592 to the dollar.

9.3 Results

9.3.1 History of Tourism at Pangandaran

One of the first documented "tourists" to the area we now know as Pangandaran was Bujangga Manik, a prince (*tohaari*) at the court of Pakuan in the west Java. At the end of the fifteenth century, he crossed the Ciputrapinggan river to arrive "at Pananjung, alongside the island *(nusa)* of Wuluheun." Pananjung is part of Pangandaran (*tanjung* meaning cape), not far east of which the Ciputrapinggan discharges itself into the ocean and Wuluheun (*wuluh,* sacret Bali bamboo *Schizostachyum brachycladum*) may be the former name of this peninsula which in the past may have been a small island off the coast (Noorduyn 1982).

Cribb (1995) gave an overview of international tourism on Java for the period 1900–1930 and noted that "most destinations had a relatively one-dimensional touristic identity." The tourists' itineraries quite naturally tended to follow the routes of least resistance, where facilities and accommodation were available (Nuryanti 1998). The Official Tourist Bureau (Officiële Vereeniging voor Touristenverkeer) offered a limited number of itineraries, for instance, for an 18-day overland trip from Jakarta via Bogor to Yogyakarta and Mt. Bromo Tengger. Day 10 brought you in the vicinity of Pangandaran (the night was spent at Maos somewhat further to the east) (Cribb 1995). Steinmann and Scheibener (1925) recommended Pangandaran to nature lovers and biologists as it was one of the most beautiful areas of the southern Java. In 1922, the railroad to Pangandaran was completed—allowing easy access—and two hotels on the isthmus provide accommodation.

In terms of international tourism and itineraries, little had changed in the 1970s when the first studies on ebony langurs were conducted in Pangandaran. After having visited the mountain areas of Bogor (Mt Gede-Pangrango), Bandung (Mt Tangkuban Perhahu) and Garut (Mt Papandayan), for many international tourists Pangandaran with its white beaches was an excellent point to recuperate before visiting Yogyakarta and the Borobudur. In 1973, some 53,000 tourists visited the isthmus and in 1978 this had grown to 136,000. After being faced with declining oil revenues in the late 1980s, the Indonesian government diversified and tourism, initially international and later also domestic, became more important (Nuryanti 1998). From 1978 onwards, we have good data on the number of tourists that visit the isthmus (Fig. 9.2). Between the early 1980s to the late 1990s, there was a slow increase in the number of tourists that visited Pangandaran. This dropped considerably when the tsunami hit, but from 2007 onwards there was a steep increase in tourist numbers (Fig. 9.2). In 2018 4.1 million people, 99.8% from within Indonesia, visited the isthmus.

Based on the number of international and domestic tourists, the lengths of their stay (trips from domestic visitors last shorter than that of foreign tourists, i.e., 2 and 4 days on average, respectively), and day of the week, Nijman (2021) estimated that on an average day in 2019 ~ 23,000 people (residents and tourists) were present on

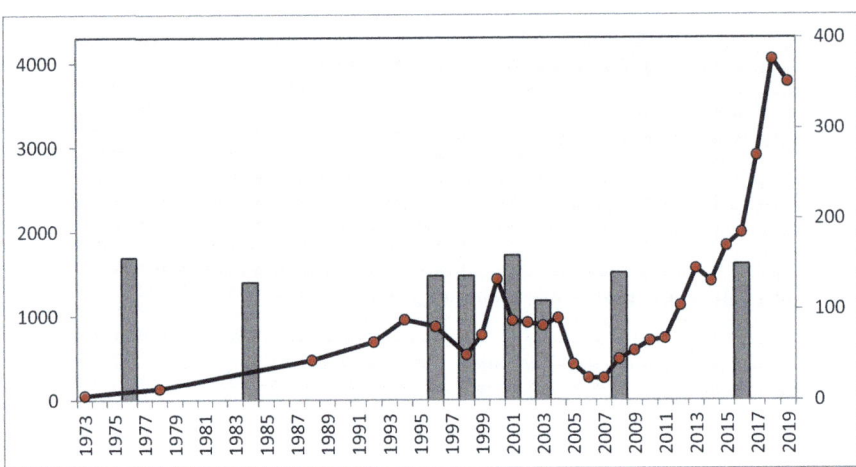

Fig. 9.2 Development of tourism in Pangandaran over the last 40+ years (line, in thousands), showing an initial gradual increase from the late 1970s to the early 2000s, followed by a dip around 2006 when a tsunami hit the peninsula, and an exponential increase since 2007. The bars are estimates of the population of ebony langurs in the Nature Tourism Park (in individuals), that have remained stable at around 150 individuals over the entire period

the isthmus. In the weekends, this went up to ~44,000 people a day, and on an average holiday (Lebaran, New Year's Day, International Kite Festival) this increased to ~59,000 people a day. Weekends are considerably more crowded than weekdays and especially during these weekend large numbers of tourists, often as part of organised groups, enter the Nature Tourism Park for a picnic or a stroll.

From the period 2000 to 2012, we have good data on both the number of visitors that purchased a ticket to the isthmus and that bought a ticket to enter the Nature Tourism Park. For these years, there is a strong correlation between the two (Pearson's R = 0.855, R^2 = 0.73, N = 13, P = 0.0002), and on average 11.9% (range 6.7–16.9%) of all visitors do indeed purchase a ticket to enter the Nature Tourism Park. For the more recent years, data are not available but even if we use the lowest percentage from the 13-year period for which we have data, i.e., 6.7%, then for the years 2018 and 2019 271,000 and 253,000 visitors entered the Nature Tourism Park, respectively.

9.3.2 Overview of Research on Ebony Langurs in Pangandaran

In 1976 to 1978 Edy Brotoisworo, working towards his PhD at Kyoto University, conducted 24 months of fieldwork on the behavioural ecology, the ranging behaviour and population dynamics of seven groups of langurs in Pangandaran (Brotoisworo 1983, 1991). The majority of his work was conducted in the Nature Tourism Park and the north-western part of the Strict Nature Reserve. He noted about the seven groups that "They were already habituated to visitors. Nevertheless, when the troops began to enter the dense forest, they became shy and my presence always upset them. After sometime they could be well habituated which made observation becoming more easy" (Brotoisworo 1983). This was followed, in 1984 and 1985, over a period of 21 months, by Karen Kool, from the University of New South Wales, who studied the behavioural ecology, feeding and food selection of two groups of ebony langurs, i.e., Group 21 in the Strict Nature Reserve and Group 3 in the Nature Tourism Park. Kool (1989) selected Pangandaran as a study site as she noted that "The Nature Tourism Park is frequented by Indonesian tourists so it was expected that *T. auratus* groups found there would be at least partially habituated to the presence of humans. Although possibly partially habituated, *T. auratus* were known not to interact directly with humans, for example, they did not accept or take food from people".

Following this, Erri Megantara, who had obtained his PhD from Kyoto University studying banded langurs *Presbytis femoralis* in Sumatra in the 1980s, studied the langurs in Pangandaran from 1994 to 1996 (Megantara 2004). Cementing the earlier links with Kyoto University, this period also saw the first Japanese researchers focussing their attention to the langurs of Pangandaran. Kunio Watanabe and Manazumi Mitani studied the langurs in the late 1990s and later the mid-2000s

(Mitani and Watanabe 2009). In 2011 Yamato Tsjiu, from Kyoto University, and collaborators from Japan and Java, started their work on the langurs of Pangandaran, again geographically focussing on the Nature Tourism Park and small areas within the Strict Nature Reserve (Tsuji et al. 2013, 2015, 2016, 2017). This work continues until this day and Tsuji is identified as the most productive researcher on ebony langurs (Aufar 2020) (Table 9.1).

My visits to Pangandaran overlapped with that of Megantara, Watanabe, Mitani and Tsuji. These visits were more focussed on the langur population as a whole, on the conservation of the langurs and their forest environment and only to a lesser degree on specific ecological or behavioural aspects of the langurs (for instance, living in coastal habitats: Nijman 2019).

The population of ebony langurs in the northernmost part of the reserve, i.e., the Nature Tourism Park and immediately adjacent areas in the Strict Nature Reserve, have remained remarkably stable over the last 40+ years. The most recent estimate from 2008 of eight groups and 140 individuals (Mitani and Watanabe 2009) is very similar to that of the first one from 1976, i.e., eight groups and 157 individuals (Brotoisworo 1983). In fact, the number of groups have been consistently estimated at between seven and nine groups (with one outlier of 12 groups: Husudo and Megantara 2002), and there was no statistical difference between these estimates ($\chi^2 = 2.10$, df = 1, P = 0.147). Only when expressed in terms of number of individuals (estimates fluctuate between 109 and 160 individuals) is the estimate from Husudo and Megantara (2002) higher than the others ($\chi^2 = 7.38$, df = 1, P = 0.007). Although qualitative, my observations also suggest a relatively stable population of ebony langurs between my first visit in 1995 and my most recent one in 2019, and the period in between.

While combined these researchers and teams have covered a very broad range of topics, resulting in one of the better-known primate populations in Indonesia, as far as I can assess hitherto none of their studies have paid detailed attention to the interaction between tourists and the langurs. Tourist do interact with langurs, but only to

Table 9.1 Non-exhaustive list of researchers that have focussed on ebony langurs in Pangandaran between 1976 and 2020

Researcher (affiliation)	Period	Journal articles[a] (citations)
Edy Brotoisworo (Kyoto University, Kyoto, Japan; Padjajaran University, Bandung, Indonesia)	1976–1978	3 (52)
Karen M Kool (University of New South Wales, Sidney, Australia)	1984–1985	4 (230)
Erri Megantara (Padjajaran University, Bandung, Indonesia)	1994–1996	4 (10)
Kunio Watanabe (Kyoto University, Kyoto, Japan), Manazumi Mitani (University of Hyogo, Hyogo, Japan)	1996–1998, 2008	6 (26)
Yamato Tsuji (Kyoto University, Kyoto, Japan), Bambang Suryobroto (Bogor Agricultural University, Bogor, Indonesia)	2011–2020	7 (42)

[a]The number of publications is indicative as some author(s) also published on related topics, including, for instance, the vegetation development within the reserve, and several authors have co-authored papers together

a limited extent. The main reason for the limited interaction is because the langurs in Pangandaran are mostly arboreal and remain high up in the trees, typically at least 10–15 m above the forest floor. Even with about a quarter of a million people entering the Nature Tourism Park, as I have experienced, many of them do not notice the langurs. During each and every visit to Pangandaran there were numerous instances where I was observing the langurs in the trees and tourists walked under them oblivious of their presence. On afternoons, just before sunset, langurs would move out of the Nature Tourism Park into the rows of sea almond trees that line the beaches up to several hundred metres onto the isthmus and into the village. Again, very rarely were the langurs noted by tourists. In contrast, the more terrestrial long-tailed macaques are difficult to miss. Photographs posted on the Internet support this notion of tourists observing and interacting more with macaques than with langurs. Of the 366 photographs linking primates to Pangandaran, 301 were long-tailed macaques and 65 were langurs. The search was conducted such that we expect an equal number, but the findings were significantly different (binominal test, $z = 12.28$, $P < 0.00001$).

9.3.3 Tourism and Tourism-Related Research in Pangandaran

The first researchers that seriously explored various aspects of tourism in Pangandaran were Wilkinson and Pratiwi (1995). They were especially interested in the gender roles and relationships, including employment patterns, income, family structure and who took care of the children within households, and how that was affected by tourism. The research was conducted in the period 1989–1992 at a time when Pangandaran was still very much a fishing village. Around two-fifth of the population were fishers and, corrected for inflation, a non-boat owning fishing family had a monthly income of ~US$142. They described the situation with regards to tourism as follows:

> There were 692,076 tourists during April 1991-March 1992; only 13,703 (1.98%) were foreign tourists. Most domestic tourists come from Bandung (60%) and Jakarta (10%), followed by other cities in West and Central Java. The major origins of foreign tourists are the Netherlands, Germany, United Kingdom, Switzerland, France, United States, and Australia. Domestic tourists stay for shorter periods of time (an average of two days) than foreign tourists (five days). No data are available on either expenditures or the gender of tourists. Observation suggests that domestic tourists tend to travel in families or groups, with the exception of some individual males; foreign tourists travel as husband-wife pairs or small groups, and individual females are not uncommon. (Wilkinson and Pratiwi 1995: 292)

This accurately describes Pangandaran when I first visited the area in 1995. Wilkinson and Pratiwi (1995) found that in the peak tourist season many fishing boat owners switch their boats from fishing to transporting tourists to the Strict Nature Reserve. In the early 1990s, tourism was already much more lucrative activity than fishing, as for a several hours long trip around the reserve, the price was US$28. Tourism at that time was already recognised as a double edge sword, in that it would bring greater economic prosperity, but that it could also lead to a dramatic change in village life. Protests emerged when it was revealed that plans were developed to erect a five-star hotel on the East Beach Road on village land that was used for cattle grazing, village festivals, fishing boat and net repair. Other plans to increase tourism were the establishment of an airport and a golf course. Wilkinson and Pratiwi (1995) noted that concerns were expressed about the pressures that such developments would place on already strained supplies of skilled personnel, services (e.g., water, sewer, electricity, roads, public transportation, housing for additional non-local workers) and the environment (e.g., groundwater, the nature reserve, loss of agricultural lands).

Rosyidie et al. (2010) explored the relationships between climate (i.e., mean monthly temperature, rainfall and humidity) and visitor numbers for the period of 1998–2008. In this period, the mean monthly temperature increased somewhat, i.e., between 0.1 and 0.5 °C, whereas visitor numbers increased (Fig. 9.1). Given these small temperature changes, it was not surprising that the majority of tourists did not feel that climate change had led to a change in attractiveness of Pangandaran as a tourism site. Instead, a third of them felt the narrowing of the beach and the decrease of vegetation (mainly along the beach front) had an impact on their tourism experience.

Dhalyana and Adiwibowo (2013) identified the type of job opportunities that were created by nature tourism in Pangandaran, analysed the income this generated and explored the influence of tourism on the social life of the local community. They found that ~700 people were employed in hotels and homestays, ~500 people were employed, at least occasionally, to provide transport (motorbike, minibus, boats, etc.), several hundred worked in restaurants and cafés, and ~700 worked on the beaches selling food, drinks, bathing essentials, etc. Only 60 people were officially employed as guides to enter the Nature Tourism Park and the Strict Nature Reserve. The income that was generated differed substantially, with, for instance, a boat taxi driver making ~US\$231 month^{-1}; a food stall operator, ~US\$352 month^{-1}; and a bicycle taxi driver ~US\$72 month^{-1}. While in general there was a high degree of cooperation between various businesses and local communities, concerns were expressed about the rising number of tourists and how changing lifestyle and crime may influence village life.

In terms of the type of research that has been conducted in Pangandaran over the last 50 years, there are some marked changes over time (Table 9.2). The tsunami and its aftermath could only be studied after 2006 and indeed in that period there was a clear emphasis on this. Papers on the primates in the reserve have been published with some regularity but ones focussing on tourism only were published in the second half of this period. Kool's doctoral thesis on the behavioural ecology of the ebony langur was the most cited work published in the 1980s and the paper by Wilkinson and Pratiwi on gender, and tourism was the most cited paper from the 1990s.

Table 9.2 Overview of English language studies conducted in Pangandaran over the last 50 years. Citations are from Google scholar up until December 2020

Period	Tourism	Physical geography, tsunami	Botany	Primates	Other animals	Fisheries	Most cited [citations]
1970–1979	0	1	2	1	3	2	Vegetation analysis [28]
1980–1989	0	0	0	4	1	4	Langur behavioural ecology [46]
1990–1999	3	0	2	7	1	7	Gender and tourism [245]
2000–2009	5	12	2	3	2	8	Tsunami [139]
2010–2019	7	3	2	5	8	6	Vegetation analysis [19]

9.4 Discussion

9.4.1 Primate Tourism on Java—Pangandaran as a Case Study

I report here on tourism and ebony langurs in Pangandaran noting that while over the last four decades tourist numbers have increased from a few hundred thousand to four million (Whitten et al. 1996; Nijman 2021), the population of langurs has remained stable. The oldest ebony langur lived in captivity until the age of nearly 32 years and indeed other *Trachypithecus* langurs have been recorded to live into their late 20s (Weigl 2005). While it is certainly not a given that ebony langurs in the wild live to a similar long age, but in the absence of any major predators in Pangandaran this is also not impossible. Brotoisworo (1983) remarked on some old males and females among the ~90 ebony langurs he studied, and it is worth considering that very old individuals he observed in the 1970s may have been born in the 1950s. Some of the older individuals that are around in 2020 may have been born in the 1990s and infants born in recent years may live into the 2050s. These individuals have seen, or will see, a lot of change in their lifetime.

Pangandaran is unique in Java in that it is the only site where one of its five primate species has been studied more or less continuously over four decades. Unfortunately, not all of the work published on the ebony langurs has reached a wide audience and even in the last 5 years (2015–2020) the one study conducted in the 1980s continues to receive more citations than all the other studies combined.

What is apparent while quantifying research that has been conducted in Pangandaran is how little overlap or interconnectedness there is between the disciplines, or even sometimes within disciplines. For instance, Kool (1989), in her 296 pages doctoral dissertation, only makes reference twice to Brotoisworo's earlier work (once to indicate that he described the area as everwet and once that he had noted that the unpigmented area in the female's pelvic region forms a distinct pattern which may aid in recognition of individuals), even though both had researched the same langur groups only 6 years apart. No reference to Brotoisworo's work was made in two other reports (Kool 1992; Kool and Croft 1992) suggesting limited continuity in the two studies. Focussing on primates and tourism, one would expect that over these four decades the primatologists would have focussed their attention on the tourists and the broader economic activities that play out in parts of the home ranges of the langurs. Conversely, with an estimated quarter of a million people visiting the Nature Tourism Park annually, where many of them must come into visual contact with the langurs, one would hope that the primates feature in any research that is conducted on the tourists and the tourism industry. However, neither of this seems to be the case. While perhaps the interconnectedness between (eco) tourism and primatology is something that could be (actively) avoided in the 1970s, 1980s and 1990s, certainly in the last decade, with increasing tourist numbers and a building boom, it becomes more apparent that this needs to be considered.

There is ample scope for promoting eco- or primate tourism in Pangandaran, especially when one makes better use of the wealth of data that is available on the langurs. For instance, while both Brotoisworo (1983) and Kool (1992) make it clear that there is no clear birth season for the langurs in Pangandaran (Kool (1992) hints as a peak in March and October), only by combining data from various researchers does it become clear that indeed infants are born in every month in equal numbers (Fig. 9.3). Infant ebony langurs are brightly yellow coloured and are easily pointed out to interesting tourists.

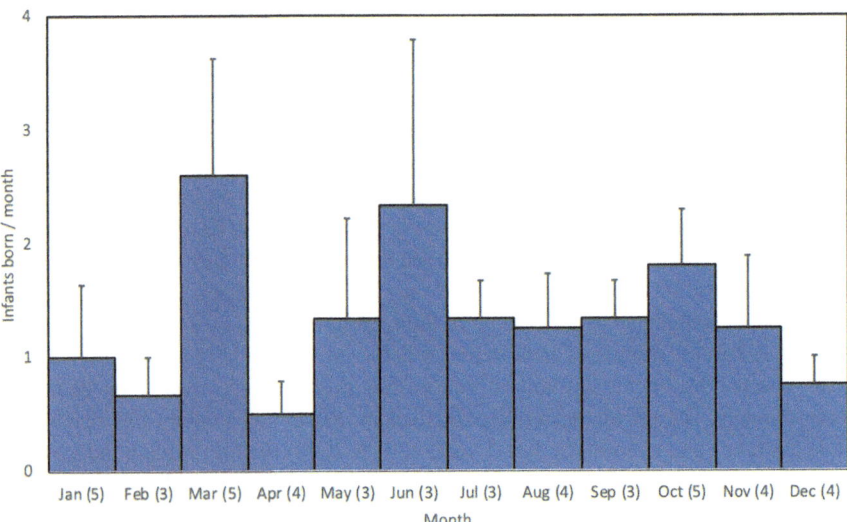

Fig. 9.3 Months during which births of ebony langurs were observed or deducted in Pangandaran (Mean ± SE), based on the data from Brotoisworo (1983), Kool (1992), Trisilo et al. (2021) and my own observations between 2004 and 2018 (62 births in total). Numbers between brackets are the number of years for which data were available

9.4.2 Conservation and Ecotourism in a Broader Perspective

The focus hitherto has been largely on the langurs in the Nature Tourism Park and how this may have been affected differently by the (increased) presence of tourists. The Pangandaran reserves comprise the Nature Tourism Park (38 ha) and the Strict Nature Reserve (419 ha), and relatively few people enter the latter area. A Strict Nature Reserve should not be open to the public, but in Pangandaran (tacitly) approved guides are present to offer tours. Even if just 1% of the tourists that enter the Nature Tourism Park also enter the Strict Nature Reserve, then annually 2500–3000 people enter it. With the exception of Kawah Ijen Merapi Ungup-ungup (Table 9.3), this is probably more than any other Strict Nature Reserve in Indonesia. Kool (1989) estimated that there were about 75 groups of ebony langurs in the Strict Nature Reserve—with an average group size of 10.3 individuals (smaller than in the Nature Tourism Park) this suggest a combined population of some 900 langurs. Many of these will see tourists only rarely. Population and group sizes of the langurs have remained remarkably unaffected by the large numbers of visitors, and one can speculate about the impact it must have on predators of the langurs. Leopards *Panthera pardus*, dhole *Cuon alpinus*, changeable hawk-eagles *Nisaetus cirrhatus*, amongst others can be significant predators of ebony langurs, but these are no longer present in Pangandaran (leopards do still occur in the hills north of Pangandaran: Whitten et al. 1996). Reticulated pythons, *Malayopython reticulatus*, have been postulated as potential predators of ebony langurs (Brotoisworo 1983; Tsuiji et al.

Table 9.3 Protected areas in Java, Indonesia where ebony langurs can be observed, with a description of the habitats and the number of visitors that frequent the park in 2019 (rounded to the nearest 100). Sites are listed from west to east; TN = Taman Nasional, National Park; TWA = Taman Wisata Alam, Nature Tourism Park, CA = Cagar alam, Strict Nature Reserve

Site, province	Habitat	Domestic visitors	International visitors	Ease of observing langurs
Halumun TN, Banten	Hill and montane forest	168,400	200	Difficult
Pancar TWA, West Java	Hill forest	57,900	100	Moderate
Gede-Pangrango TN, West Java	Montane forest	348,500	1400	Difficult
Pangandaran TWA/CA, West Java	Lowland and coastal forest	263,000	1900	Easy
Merapi, TN, Central Java/ Yogyakarta	Montane forest	148,300	200	Difficult
Grojogan Sewu, Lawu, TWA, Central Java	Montane forest	344,900	400	Difficult
Kawah Ijen Merapi Ungup-ungup, CA, East Java	Montane forest	126,400	25,100	Moderate
Baluran TN, East Java	Lowland and coastal forest	245,000	1400	Easy

2016), and these do occur in all parts of the forest. Brotoisworo (1983) made a convincing case that village dogs were, in fact, the most effective predator of ebony langurs, but if this is indeed still the case, then this would only affect those groups that roam closest to the isthmus.

Besides the langurs, there is (or was) a wealth of other wildlife present in and around Pangandaran, terrestrial and marine. Some of these species have been much more negatively affected by tourism and the need for souvenirs (Fig. 9.4). In the 1980s and 1990s on the isthmus, many souvenir shops sold stuffed animals, including Asian palm civets *Paradoxurus hermaphroditus*, Sunda leopard cats *Prionailurus javanensis*, water monitor lizards *Varanus salvator*, olive ridley turtles *Lepidochelys olivacea* and hawksbill turtles *Eretmochelys imbricata*, chambered nautilus *Nautilus pompilius* and horned helmet *Cassis cornuta* shells, and blown-up and dried puffer fish (Nijman 2015a, b). In the 2000s, fewer stuffed turtles and civets were present, and in the 2010s, it is mainly marine mollusc shells, pufferfish and a small number of stuffed turtles that are openly on display (Nijman 2019; Nijman et al. 2016; Nuryanto and Bhagawati 2020). Protected species of fish, including at least occasionally whale shark *Rhincodon typus*, manta ray *Mobula alfredi* and *M. birostris*, hammerhead sharks *Sphyrna* spp. continue to be traded (Hernawati et al. 2018; V. Nijman unpubl. data), and with a greater number of tourists present now than

Fig. 9.4 Tourism interactions with wildlife in Pangandaran, from top left, clockwise: Tourist sitting on a legally protected whale shark *Rhincodon typus* (2011); long-tailed macaques *Macaca fascicularis* entering a food stall on the beach (2015); legally protected green turtle *Chelonia myda* and hawksbill turtles *Eretmochelys imbricata* offered openly for sale (2015); Javan deer *Rusa timorensis* eating garbage at the coach parking lot (2015)

40 years ago, the negative impact of this on already imperilled populations most likely has increased.

The macaques and deer now feed extensively on garbage left by tourists, but the effects of this on the animals has yet to be studied. There is no data to support that any of the tourism activities in Pangandaran follow the main basic tenets of ecotourism (uniting biodiversity conservation, communities, and sustainable travel; building environmental awareness and respect; provide direct financial benefits for conservation) other than generating financial benefits for local people. At best, when focussing just on the langurs, the walk into the Nature Tourism Park and/or the Strict Nature Reserve can be described as a short ecotourism experience as part of a multi-purpose trip. The long-term presence of the langurs with no noticeable changes in their population numbers and very limited physical interaction between humans and langurs supports the conclusion that the tourists have left the area in the same condition as when they arrived. This conclusion, however, is not true for other wildlife.

Acknowledgements I would like to thank Angela Achorn, Sharon Gursky and Jatna Supriatna for inviting me to contribute to this volume, thus allowing me to put the data I collected over 25 years into the context. Various people accompanied me on my visits to Pangandaran or provided details on conservation practises in the area, and my sincere thanks to Anna Nekaris, Edo Govertse, Maartje Hilterman, Elizabeth Burgess, Denise Spaan, Pak Budin and Abdullah Langgeng. Also, I would like to thank the reviewer for constructive comments and suggestions for the improvement over this chapter.

References

Aufar FRM (2020) Studi literatur dan bibliometrik tentang Lutung Jawa (*Trachypithecus auratus*). (Unpublished BSc dissertation), Institut Pertanian Bogor, Bogor

Brotoisworo E (1983) Population dynamic of lutung (*Presbytis cristata*) in Pananjung-Pangandaran nature reserve, West Java. In: Training course on wildlife ecology. Bogor, Biotrop, pp 1–24

Brotoisworo E (1991) The lutung (*Presbytis cristata*) in Pananjung Pangandaran Nautre reserve—social adaptation to space. Comp Primatol Monogr 3:45–148

Cochrane J (1996) The sustainability of ecotourism in Indonesia: fact and fiction. In: Environmental change in South-East Asia: people, politics and sustainable development. Routledge, Abingdon, pp 237–259

Cribb R (1995) International tourism in Java, 1900–1930. South East Asia Res 3:193–204

Dalem AAGR (2002) Ecotourism in Indonesia. Linking green productivity to ecotourism: experiences in the Asia-Pacific region. Asian Productivity Organisation, Tokyo

Dhalyana D, Adiwibowo S (2013) Influencing nature tourism park of Pangandaran to the social life of local communities. Sodality: Jurnal Sosiologi Pedesaan 1:182–199

Ghimire KB (2013) The native tourist: Mass tourism within developing countries. Routledge, Abingdon.

Gunawan MP (1996) Domestic tourism in Indonesia. Tour Recreat Res 21:65–69

Hernawati D, Amin M, Irawati MH, Indriwati SE, Chaidir DM, Meylani V (2018) Potential, production and management recommendation of shark and ray in the Pangandaran area – West Java. Prosiding Simposium Nasional Hiu Pari Indonesia 2:285–291

Husudo T, Megantara EN (2002) Distribusi dan daerah jelajah lutung (*Trachypithecus auratus sondaicus*) di Taman Wisata Alam Pangandaran. J Biotika 1:36–47

Kinnaird MF, O'Brien TG (1996) Ecotourism in the Tangkoko DuaSudara nature reserve: opening Pandora's box? Oryx 30:65–73

Kool KM (1989) Behavioural ecology of the silver leaf monkey, *Trachypithecus auratus sondaicus*, in the Pandangaran Nature Reserve, West Java, Indonesia (Unpublished Ph.D. dissertation). University of New South Wales, Sidney

Kool KM (1992) Food selection by the silver leaf monkey, *Trachypithecus auratus sondaicus*, in relation to plant chemistry. Oecologia 90:527–533

Kool KM (1993) The diet and feeding behavior of the silver leaf monkey (*Trachypithecus auratus sondaicus*) in Indonesia. Int J Primatol 14:667–700

Kool KM, Croft DB (1992) Estimators for home range areas of arboreal colobine monkeys. Folia Primatol 58:210–214

Megantara EN (2004) Distribution and population of lutung (*Trachypithecus auratus sondaicus*) in Pangandaran Natural Reserve (Indonesia). Jurnal Ilmu-ilmu Hayati dan Fisik 6:260–271

Mitani M, Watanabe K (2009) The situation of the Pangandaran Nature Reserve in West Java, Indonesia in 2008, with special reference to vegetation and the population dynamics of primates. Prim Res 25:5–13

Moher D, Liberati A, Tetzlaff J, Altman DG, Prisma Group (2009) Preferred reporting items for systematic reviews and meta-analyses: the PRISMA statement. PLoS Med 6:1000097

Nakashima K (2001) Ecotourism action plan of Gunung Halimun National Park. Biodiversity Conservation Project-JICA, Bogor

Nijman V (2000) Geographic distribution of ebony leaf monkey *Trachypithecus auratus* (E. Geoffroy Saint-Hilaire, 1812) (Mammalia: Primates: Cercopithecidae). Contrib Zool 69:157–177

Nijman V (2013) One hundred years of solitude: effects of long-term forest fragmentation on the primate community of Java, Indonesia. In: Marsh LK, Chapman CA (eds) Primates in fragments: complexity and resilience. Springer, New York, pp 33–45

Nijman V (2015a) Distribution and ecology of the most tropical of the high-elevation montane colobines: the ebony langur on Java. In: Grow NB, Gursky-Doyen S, Krzton A (eds) High altitude primates. Springer, New York, pp 115–132

Nijman V (2015b) Decades-long open trade in protected marine turtles along Java's south coast. Mar Turt Newsl 144:10–13

Nijman V (2019) Ebony langurs in mangrove and beach forests of Java, Bali and Lombok. In: Nowak K (ed) Primates in flooded habitats. Cambridge University Press, Cambridge, pp 99–104

Nijman V (2021) Tourism developments increase tsunami disaster risk in Pangandaran, West Java, Indonesia. Int J Disaster Risk Sci 12:764–769

Nijman V, Spaan D, Sigaud M, Nekaris KAI (2016) Addressing the open illegal trade in large marine mollusc shells in Pangandaran, Indonesia. J Ind Nat Hist 4:12–18

Noorduyn J (1982) Bujangga Manik's journeys through Java; topographical data from an old Sundanese source. Bijdragen tot de Taal-, Land- en Volkenkunde 138:413–442

Nuryanti W (1998) Tourism and regional imbalances: the case of Java. Indonesia Malay World 26:136–144

Nuryanto A, Bhargawati D (2020) Evaluation of conservation and trade status of marine ornamental fish harvested from Pangandaran coastal waters, West Java, Indonesia. Biodiversitas 21:512–520

Pangemanan A, Maryunani LH, Polii B (2012) Economic analysis of Bunaken Nasional Park ecotourism area based on the carrying capacity and visitation level. Asian Trans Basic Appl Sci 2:34–40

Reese S, Cousins WJ, Power WL, Palmer NG, Tejakusuma IG, Nugrahadi S (2007) Tsunami vulnerability of buildings and people in South Java? Field observations after the July 2006 Java tsunami. Nat Hazards Earth Syst Sci 7:573–589

Ross S, Wall G (1999) Evaluating ecotourism: the case of North Sulawesi, Indonesia. Tour Manag 20:673–682

Rosyidie A, Adriyani Y, Suwarto T (2010) The influence of climate factors on tourist visits in Pangandaran coastal tourism area. Asian J Hosp Tour 9:87–100

Siburian R (2006) Pengelolaan Taman Nasional Gunung Leuser Bagian Bukit Lawang Berbasis Ekowisata. Jurnal Masyarakat dan Budaya 8:67–90

Steinmann A, Scheibener E (1925) Zwerftochten over het schiereiland Penandjoeng. De Tropische Natuur 14:49–56

Trisilo SP, Widayati KA, Tsuji Y (2021) Effect of infant pelage colour on infant caring by other group members: a case study of wild Javan lutungs (*Trachypithecus auratus*). Behaviour 158:277–290

Tsuji Y, Widayati KA, Hadi I, Suryobroto B, Watanabe K (2013) Identification of individual adult female Javan lutungs (*Trachypithecus auratus sondaicus*) by using patterns of dark pigmentation in the pubic area. Primates 54:27–31

Tsuji Y, Widayati KA, Nila S, Hadi I, Suryobroto B, Watanabe K (2015) "Deer" friends: feeding associations between colobine monkeys and deer. J Mammal 96:1152–1161

Tsuji Y, Prayitno B, Suryobroto B (2016) Report on the observed response of Javan lutungs (*Trachypithecus auratus mauritius*) upon encountering a reticulated python (*Python reticulatus*). Primates 57:149–153

Tsuji Y, Ningsih JIDP, Kitamura S, Widayati KA, Suryobroto B (2017) Neglected seed dispersers: endozoochory by Javan lutungs (*Trachypithecus auratus*) in Indonesia. Biotropica 49:539–545

Weaver DB (2001) Ecotourism as mass tourism: contradiction or reality? Cornell Hotel Restaur Admin Q 42:104–112

Weigl R (2005) Longevity of mammals in captivity; from the living collections of the world. Kleine Senckenberg-Reihe, Stuttgart

Whitten AJ, Soeriaatmadja RE, Afiff SA (1996) The ecology of Java and Bali. Periplus, Singapore

Widiastuti W (2020) Domestic tourism in Indonesia: another story of inequality between Java and non-Java. J Ind Tour Dev Stud 8:45–49

Wilkinson PF, Pratiwi W (1995) Gender and tourism in an Indonesian village. Ann Tour Res 22:283–299

Chapter 10
Indigenous Bird Ecotourism in Halmahera Island, Indonesia

M. Nasir Tamalene, Akhmad David Kurnia Putra, Ericka Darmawan, Mustafa Mansur, and Bahtiar

Abstract Bird watching hobbyists will often go to great lengths to observe wild birds around the world, traveling to tropical rainforests, beaches, and even mountains. As birds may serve as cultural symbols in which their songs and other sounds are important in various activities, local communities can use a cultural approach to protect birds in close proximity. The results of this study show that endemic bird species are used as cultural symbols by indigenous people on the Indonesian island of Halmahera, a tourist destination for local and foreign visitors. The study used a random survey method which involved interviewing farmers in four regions: Loloda Kepulauan, Maba, Buli, and Wangongira, as well as research respondents residing in forest fringes around the village. Interviews were conducted in local languages. The research findings reveal that the endemic birds most sought after by photographers and tourists belong to eight families: *Paradisaeidae*, *Alcedinidae*, *Pittidae*, *Rallidae*, *Megapodiidae*, *Columbidae*, *Aegothelidae*, and *Meliphagidae*. Furthermore, this study demonstrates how indigenous knowledge can be used to protect local birds by making a species a cultural symbol. These results emphasize the importance of building partnerships with indigenous communities and will hopefully encourage government programs to increase the role of local communities in biodiversity conservation. An ecotourism approach based on indigenous

M. N. Tamalene (✉)
Ethnobiological, Universitas Khairun, Ternate, Indonesia

A. D. Kurnia Putra
National Park Aketajawe Lolobata, Maluku Utara, Indonesia, Ternate, Indonesia

E. Darmawan
Biologi Education Universitas Tidar, Magelang, Central Java, Indonesia

M. Mansur
Tourism Study Program Universitas Khairun, Ternate, Indonesia

Bahtiar
Biologi Education Universitas Khairun, Ternate, Indonesia

© The Author(s), under exclusive license to Springer Nature Switzerland AG 2022
S. L. Gursky et al. (eds.), *Ecotourism and Indonesia's Primates*, Developments
in Primatology: Progress and Prospects, https://doi.org/10.1007/978-3-031-14919-1_10

knowledge is the key to sustainable development as it combines ecological, economic, and cultural dimensions. Finally, the involvement of women in ecotourism may be especially important, based on the evidence that women play a significant role in conservation activities in our study communities.

Keywords Ecotourism · Indigenous · Birds · Culture

10.1 Introduction

Reckless behaviors of tourists towards birds has can drastically alter species' natural behavior (Hakim 2017; Wolf et al. 2019). However, properly managed bird watching offers one of the most positive incentives for protected areas in highly biodiverse regions of the world: the potential to improve both the local economies and foreign exchange in the tourism sector (Steven and Jones 2014). In cases where bird watching tours do not systematically influence the ecological conditions of these species in natural habitats, continuous bird watching is likely to influence the species' behavior. Therefore, understanding the indigenous community's knowledge in conserving birds is necessary and crucial information for tourists. This study aims to demonstrate the importance of a cultural approach to bird conservation, as bird watchers, photographers, and other tourists visiting Halmahera Island, Indonesia, may be unfamiliar with this framework.

The main goal of bird watching is typically to observe birds in their natural habitat. In pursuit of this goal, it is extremely important that tourists do not harm or endanger the birds. According to bird data in Indonesia, 1769, 1771, and 1777 species were found in 2017, 2018, and 2019, respectively. Meanwhile, in 2020, Indonesian birds released the latest data on bird status, 1794 species. Biodiversity in the form of Indonesian bird species is bound to have both positive and negative impacts on ecology. In addition, bird watching is a thriving specialty market, as Europeans are increasingly attracted to the diversity of these species, with a variety of beautiful body colors, in a wide variety of habitats, from beaches to mountains.

Furthermore, interest in tourism in the wild continues to increase, and people with the hoppy always look for opportunities to see birds in the wild. However, the knowledge of tourists on native birds and community culture in traditional bird conservation needs to be known from an early age, to ensure the existence of tourists does not interfere with bird activity in natural habitats. This requires tourists' knowledge about native birds on Halmahera Island as well as the knowledge of indigenous communities in using birds as cultural symbols. Therefore, in this paper, information about bird species on Halmahera Island as a tourist destination as well as knowledge of indigenous communities in bird conservation based on local culture are provided.

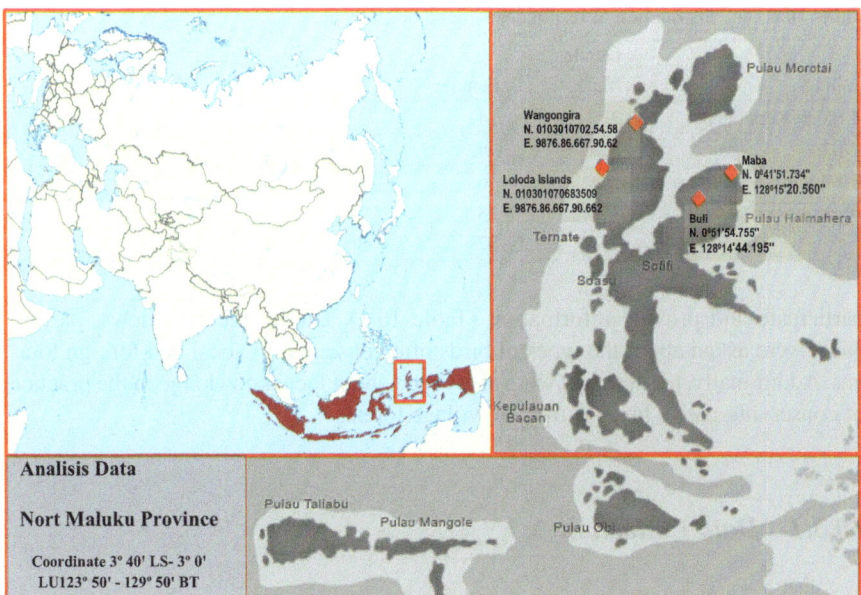

Fig. 10.1   The study area

10.2   Methods

10.2.1   Research Location

The selected study area was Halmahera Island, Indonesia (Fig. 10.1), the largest of the Maluku Islands, with a land area of 17,780 km^2. This is a tropical area with an average annual temperature of 23–31°C and an average humidity of 60–90%. The survey was conducted in the western, eastern, northern, central, and southern parts of Halmahera Island. Here, 90% of the community lives in coastal areas and works as farmers cultivating rice fields near the primary forest's edge.

10.3   Interview

Villages' local farmers were targeted as respondents for this study as they have a better understanding of the environment's characteristics, especially bird habitats in rural forests. Random visits were made to farmers in four areas on Halmahera Island, with a focus on those living near the forest's edge. Interviews were conducted in local languages for better understanding, and questions were asked thrice to ensure the validity of the respondents' answers. Out of the 44 residents visited at the time of data collection, 36 respondents (27 female and 9 male) agreed to

Table 10.1 The distribution of respondents based on study area

Village	Gender		No. of persons	Percentage
	Male	Female		
Loloda Kepulauan	2	14	16	44.44
Maba	2	4	6	16.67
Buli	2	6	8	22.22
Wangongira	3	3	6	16.67
	9	27	36	100.00

participate and provide information (Table 10.1). During the interviews, participants were asked about the types of birds often observed by local and foreign tourists. Additionally, participants were also asked about local knowledge in the practice of conserving native birds within the village forest.

10.3.1 Data Collection

The main fieldwork was conducted by the authors from March to December 2020. Data collection included participant observation, semi-structured interviews, and household surveys (Mann 2016). The interviews were conducted from afternoon until evening, after farmers came home from the gardens. These interviews took place in houses near gardens when available, at which point bird species around the gardens and forests were observed. Recording tools and transcript books of field notes were used to record all information from the local community. Bird species data were identified using the *Burungnesia* Indonesia application, as well as https://www.burung.org/ and https://www.iucnredlist.org/ for endemic and endangered taxa, respectively. The data were then transferred to a computer device for analysis.

10.3.2 Data Analysis

This study used a qualitative thematic data analysis (Chiwanga and Mkiramweni 2019). The meaning of each wild bird species' sound is explained using the knowledge of indigenous communities. Thus, each species is assessed according to the importance attributed by respondents and does not depend on the researcher's judgement.

10.4   Analisis Data

10.5   Results

10.5.1   Endemic Bird Species as Tourist Destinations on Halmahera Island

The endemic birds on Halmahera Island, Indonesia, that are most sought after by photographers and tourists belong to eight families: *Paradisaeidae, Alcedinidae, Pittidae, Rallidae, Megapodiidae, Columbidae, Aegothelidae,* and *Meliphagidae* (Figs. 10.2, 10.3, 10.4, 10.5, 10.6, 10.7, and 10.8).

10.5.1.1   Types of Birds in Paradisaeidae Family

Wallace's standardwing (Indonesian name, *Bidadari Halmahera*; scientific name, *Semioptera wallacii* (G. R. Gray, 1859)) is a bird-of-paradise belonging to the *Cendrawasih* family. This species can be observed at the following locations: Tayawi and Ake Jawi Resorts in Aketajawe Lolobata National Park, Foli Village, Wasile District, East Halmahera, Weda Resort in Kobe Village, Central Halmahera, and Pasir Putih in the West Halmahera Regency. Another bird from the same family, the paradise crow (Indonesian name, *Cendrawasih*; scientific name *Lycocorax pyrrhopterus* (Bonaparte, 1850)) is also found on the island. It is common in all forest and plantation areas within Halmahera. Importantly, cendrawasih found on Halmahera Island differ slightly in appearance from those found on Obi Island.

Fig. 10.2 (a) *Semioptera wallacii* (G. R. Gray, 1859) (Wallace's standardwings Bird of Paradise), (b) *Lycocorax pyrrhopterus.* (Bonaparte, 1850)

Fig. 10.3 (**a**) *Todiramphus diops* (Temminck, 1824), (**b**) *Todiramphus funebris* (Bonaparte, 1850), (**c**) *Tanysiptera galatea* (G. R. Gray, 1859), (**d**) *Ceyx azureus* (Latham, 1801), (**e**) *Ceyx azureus*. (Latham, 1801)

Fig. 10.4 (**a**) *Pitta maxima* (S. Muller & Schlegel, 1845), (**b**) *Erythropitta erythrogaster ruviventris*. (Heine, 1860)

Fig. 10.5 *Habroptila wallacii*. (Gray, 1860)

Fig. 10.6 *Habroptila wallacii*. (Gray, 1860)

10.5.1.2   Types of Birds in Alcedinidae Family

The blue-and-white kingfisher (Indonesian name, *Cekakak biru-putih*; scientific name, *Todiramphus diops* (Temminck, 1824)) is endemic to the North Maluku. After it rains, this bird can often be observed perching on dry branches or on electrical cables on the highway.

The Sombre kingfisher (Indonesian name, *Cekakak murung*; scientific name, *Todiramphus funebris* (Bonaparte, 1850)) is typically found in one or more pairs. This species is rarely seen, which makes it especially sought after by bird watchers.

Fig. 10.7 *Ptilinopus monacha.* (Temminck, 1824)

Fig. 10.8 *Aegotheles crinifrons.* (Bonaparte, 1850)

In contrast to the Sombre kingfisher, the common paradise kingfisher (Indonesian name, *Cekakak-pita*; scientific name, *Tanysiptera galatea* (G. R. Gray, 1859) is frequently seen in Halmahera. Although it is not an endemic species, documentation of this racket-tailed bird is still in great demand by bird watchers and photographers. The bird is most easily found in the Aketajawe Lolobata National Park, Ake Jawi Resort, Maluku, North Maluku, and Papua.

The azure kingfisher (Indonesian name, *Raja-udang Biru-langit*; scientific name, *Ceyx azureus* (Latham, 1801)) is highly sensitive to human presence and is therefore only distributed throughout the North Malaku and Papua. However, it is very easily found in small, unspoiled rivers within the Ake Jawi Resort area of Aketajawe Lolobata National Park.

The Halmahera dwarf kingfisher, also known as the variable dwarf kingfisher (Indonesian name, *Udang-merah kerdil*; scientific name, *Ceyx Lepidus* (Temminck, 1836)), is often found in small rivers or other wetlands on the island of Halmahera, in Ternate, or throughout the North Maluku. Despite its very small size, the species is easily found at Resort Ake Jawi, where it is a favorite of wildlife photographers.

10.5.1.3 Types of Birds in the Pittidae Family

The ivory-breasted pitta (Indonesian name, *Paok Halmahera*; scientific name, *Pitta maxima* (S. Muller & Schlegel, 1845)) is scattered throughout the Halmahera Forest. The species has a distinct sound that is very easy to hear; however, it can be difficult to spot this bird due to its sensitivity to human presence. For this reason, it highly desired by bird watchers and photographers. Places like Ake Jawi Resort at Aketajawe Lolobata National Park have created observation areas to spot these birds. The largest Paok species is often captured with camouflage nets, which means watchers and photographers are unable to see the birds' activity.

The north Moluccan pitts (Indonesian name, *Paok mopo Maluku Utara;* scientific name, *Erythropitta erythrogaster ruviventris* (Heine, 1860)) is another highly sought after by bird watchers often found in Ake Jawi Resort, Weda Resort, and Sidangoli.

10.5.1.4 Types of Birds in Rallidae Family

The drummer rail (Indonesian name, *Mandar gendang*; scientific name, *Habroptila wallacii* (Gray, 1860)) is perhaps the most difficult to find by bird watchers and photographers because of its vulnerable status (cite IUCN status). This ground bird is found at Ake Jawi Resort at Aketajawe Lolobata National Park, Halmahera Island.

10.5.1.5 Types of Birds in Megapodiidae Family

The Moluccan scrubfowl (Indonesian name, *Gosong Maluku*; scientific name, *Eulipoa wallacii* (Gray, 1860)) can be seen in Simao Village, Galela, North Halmahera Regency. This area is a common tourist destination for observing endemic birds of Maluku and North Maluku. The birds are typically seen at night, as they spend the daytime hours laying eggs together on the black sand beach (communal nesters).

10.5.1.6   Types of Birds in Columbidae Family

The blue-capped fruit dove (Indonesian name, *Walik topi biru*; scientific name, *Ptilinopus monacha* (Temminck, 1824)) is the smallest walik bird of the dove-pigeon family (the *Columbidae* family) in the North Maluku. It is often found throughout the North Maluku, especially in tree canopies with small fruit (e.g., banyan trees).

10.5.1.7   Types of Birds in Aegothelidae Family

Halmahera is the only place in Indonesia where the *Aegothelidae* family is found. One species belonging to this family, the Moluccan owlet-nightjar (Indonesian name, *Atoku Maluku*; scientific name, *Aegotheles crinifrons* (Bonaparte, 1850)), is a crepuscular bird endemic to Halmahera. Our findings suggest that this species is not typically of high priority to bird watchers despite its uniqueness and endemic status. The face of the Moluccan owlet-nightjar resembles that of frogmouths from the *Podargidae* family, which is scattered throughout Indonesiaand often found at night in the Paruh Bengkok Sanctuary, Ake Jawi Resort, and forest areas in Halmahera.

10.5.1.8   Types of Birds in Meliphagidae Family (Fig. 10.9)

The white-streaked friarbird, also known as the Halmahera friarbird (Indonesian name, *Cikukua Halmahera*; scientific name, *Melitograis gilolensis* (Bonaparte, 1850)), is spread across the islands of Halmahera, Morotai, Kasiruta, and Bacan. This species is very rarely found by tourists, which may add to its desirability.

10.5.1.9   Types of Birds in the Psittaculidae Family

The gathering lorry (Indonesian name, *Nuri Ternate*; scientific name, *Lorius garrulous* (Linnaeus, 1758)) is an endemic bird found in the North Maluku province. The species is found on the Halmahera, Widi, and Ternate Islands, where it can be detected by its loud voice and tendency to fly in groups (Fig. 10.10).

Fig. 10.9 *Melitograis gilolensis.* (Bonaparte, 1850)

Fig. 10.10 *Lorius garrulous.* (Linnaeus, 1758)

Fig. 10.11 (**a**) Symbol of the Loloda Sultanate (**b**) *Cacatua alba.* (Müller, 1776)

10.5.2 Indigenous Knowledge in Cultural-Based Bird Conservation

Bird sounds have a cultural meaning which can influence the conservation of bird species. Local people often catch and care for birds with loud sounds and "crooked" beaks. For instance, the white cockatoo (*Cacatua alba*) (Müller, 1776) (Fig. 10.11) is a species endemic to Halmahera that functions as a cultural symbol in the Loloda sultanate of the North Maluku province. Several other types of birds are also used by the local community as indicators during life activities, where bird sounds have both the cultural and ecological significance. Table 10.2 presents the cultural and/or ecological meanings of various bird sounds according to Halmahera indigenous knowledge.

Returning to example of the white cockatoo (*Cacatua alba*), people in the Loloda sultanate of Halmahera Island made this bird a cultural symbol because it is respected as a sacred animal. Its white and clean feathers are believed to be a symbol of holiness, thereby requiring protection is under the concept of ***ngara mabeno***. Etymologically, the ***ngara mabeno*** translates to "door wall" (*ngara* = door, *mabeno* = wall). This term refers to protection by all indigenous peoples under the Loloda Sultanate. Meanwhile, Limau Tolimadu is Mount Loloda as a habitat for birds requiring protection. In addition to being a sacred animal, the white cockatoo is believed to be a bearer of good news. This philosophy is a common meaning symbolized by the White Cockatoo with the local name *Gatala Bobudo*, and this brings public order implemented as *Adat se-Atorang*. The topknot on the bird symbolizes the leader and people. In this study, there were no prohibitions or customary laws used as guidelines, in the community's social life.

Table 10.2 The meaning of the sound/song from the types of birds, based on a cultural approach

Family	Scientific name	English	Indonesian name	Meaning of bird sounds	Conservation status	IUCN red list of threatened species
Alcedinidae	*Alcedo azurea* (Latham, 1801)	Azure Kingfisher	*Raja Udang Biru-Langit*	To the Maba community in the eastern part of Halmahera Island, the sound of this bird signals the presence of a predator around the garden, such as an eagle, that threatens to catch the community's chickens	Not protected	Least concern (LC)
	Tanysiptera galatea (Gray, 1859)	Common Paradise-Kingfisher	*Cekakak pitta biasa*	This bird's sound is a sign that the rain has stopped	Not protected	Least concern (LC)
	Todiramphus funebris (Bonaparte, 1850)	Sombre Kingfisher	*Cekakak murung*	This bird's sound is a sign that the rain has stopped	Not protected	Least concern (LC)
	Todiramphus diops (Lesson, 1827)	Blue-and-white Kingfisher	*Cekakak biru putih*	This bird's sound is a sign that the rain has stopped	Not protected	Vulnerable (VU)
	Ceyx azureus (Latham, 1801)	Azure Kingfisher	*Raja-udang biru-langit*	This bird's sound is a sign that the rain has stopped	Not protected	Least concern (LC)
Cacatuidae	*Cacatua alba* (Müller, 1776)	White Cockatoo	*Kakatua Putih***	This bird's sound serves as an alarm to wake the farmers	Protected	Endangered (EN)

(continued)

Table 10.2 (continued)

Family	Scientific name	English	Indonesian name	Meaning of bird sounds	Conservation status	IUCN red list of threatened species
Columbidae	*Ptilinopus monachal* (Temminck, 1824)	Blue-capped Fruit-Dove	*Walik topi biru**	To the people in the central part of Halmahera Island, th sound of this bird represents wild animals	Not protected	Near threatened (NT)
Megapodiidae	*Habroptila wallacii* (Gray, 1860)	Moluccan Scrubfowl	*Gosong Maluku***	This bird's sound signifies hard work	Protected	Vulnerable (VU)
Meliphagidae	*Melitograis gilolensis* (Bonaparte, 1850)	White-streaked Friarbird	*Cikukua Halmahera*	To the Tobelo ethnic community in the northern part of Halmahera Island, the sound of this bird indicates sunny weather	Not protected	Least concern (LC)
Paradisaeidae	*Semioptera wallacii* (Gould, 1859)	Standardwing paradise	*Burung Bidadari*	This bird's sound is a sign that it is time to work in the garden	Protected	Least concern (LC)
	Lycocorax pyrrhopterus (Bonaparte, 1850)	Halmahera Paradise-crow	*Cendrawasih gagak*	This bird's sound serves as an alarm in the morning	Protected	Least concern (LC)

(continued)

Table 10.2 (continued)

Family	Scientific name	English	Indonesian name	Meaning of bird sounds	Conservation status	IUCN red list of threatened species
Pittidae	*Pitta Maxima* (Müller & Schlegel, 1845)	Ivory-breasted Pitta	*Paok Halmahera*	This bird's sound is an alarm in the morning	Protected	Least Concern (LC)
	Erythropitta erythrogaster ruviventris (Bonaparte, 1854)	Halmahera Red-bellied Pitta	*Paok mopo Maluku utara*	This bird's sound conveys two messages: that it is time to worship or that one should not be allowed to go to sea	Not protected	Least Concern (LC)
Psittaculidae	*Lorius garrulus garrulus* (Linnaeus, 1758)	Chattering Lory (nominate)	*Nuri Ternate***	This bird's sound signifies the ripeness of fruit in the garden	Protected	Vulnerable (VU)
Rallidae	*Habroptila wallacii* (Gray, GR, 1861)	Drummer Rail	*Mandar Gendang*	This bird's sound serves as an alarm to wake up farmers	Protected	Vulnerable (VU)

*consumed
**maintained & traded

10.6   Discussion

Birds on the Halmahera Island have become objects of ecotourism and important cultural indicators for indigenous people. In this study, local community members, especially women, are interested in developing ecotourism to fulfill the needs of families. According to Tran and Walter (2014), the complexity of integrating gender perspectives into community-based ecotourism provides positive benefits for local women such as women provide home stay services, make local souvenirs for tourists, and become guides. Furthermore, birds are ecologically linked to cultural values and therefore attract numerous tourists. Bird watching tourism can help promote bird conservation and can foster connections between individuals, groups, and the environment (Chiwanga and Mkiramweni 2019). Thus, the combination of indigenous and scientific knowledge helps effective conservation of biodiversity (Su et al. 2020).

Properly managed bird watching tourism also has a positive impact on protected areas in many of the world's most diverse regions, which is why it is often considered to be a sustainable tourism activity (Steven and Jones 2014). To ensure that bird watching tourism remains sustainable, future wildlife tourism ought to improve management through establishing guidelines to minimize impacts, enforcing those guidelines for all visitors and tour guides, promoting long-term wildlife monitoring and research programs, and improving education opportunities for visitors (D'Amico and M. 2017).

Bird watching tourism is one of the most popular nature-based tourism activities and, therefore, has a lot of potential to benefit those involved. Thus, maximizing the potential for management by local communities helps to empower indigenous people (Markwell 2018). The ecotourism program also maintains the natural environment's principles of local empowerment and sustainable management. Therefore, this program has the potential to benefit local communities, biodiversity, as well as visitors (Nuckel 2019). A study by Park et al. (2019) showed tourism objects, for instance, natural attractions of wild animals, natural scenery, and culture of the local community are objects for tourists visiting an area. Ecotourism increases the income of local villagers' and makes management more effective with an indigenous cultural approach (Walter 2020). Furthermore, community-based ecotourism development in the regions maximizes socioeconomic benefits and protects the both environmental and cultural resources (Zuniga 2019). Additionally, strategy development helps to promote bird-based tourism in order to increase community incomes at the local level (Maldonado et al. 2018; Stronza et al. 2019; Harbor and Hunt 2021). The prospect of developing cultural-based ecotourism is therefore the key to improving the community's economy at the village level. However, there is a need to provide tourists with knowledge of wild animal behavior to ensure that visitors are careful in ecotourism areas.

However, traveling by entering the forest needs to be accompanied by a guide and a map of a forest area that is safe to visit, because certain animals can harm visitors, such as bees, ants, plant species that have leaf structures that itch to the touch, and other wild animals. A safe forest map is the key in visiting the tourist forest area.

Threats to natural tourism attractions, such as wild birds, must be considered. For instance, improper behavior by tourists—such as disrespect towards birds and ecotourism managers, violations of the indigenous community's ethical code, or even recklessness—could be cause for concern (Maccarthy and Mary 2020). Ecotourism programs need to emphasize all the aspects of people's welfare to ensure active participation of local communities (Kibria et al. 2020). In addition, increasing tourism in sensitive natural areas without proper planning and management threatens the ecosystem and the local culture's integrity (Idris et al. 2019). According to Balasubramaniam et al. (2020), anthropogenic factors influence wildlife behavior; the "ethnoconservation" approach is therefore used as an effort to anticipate threats of indigenous knowledge loss in bird conservation. Ethnoconservation refers to the process of utilizing, protecting, managing, and preserving wisdom, by a certain community or ethnic group, through a cultural or religious approach. The method is

Fig. 10.12 Ethnoconservation values in the ecotourism development

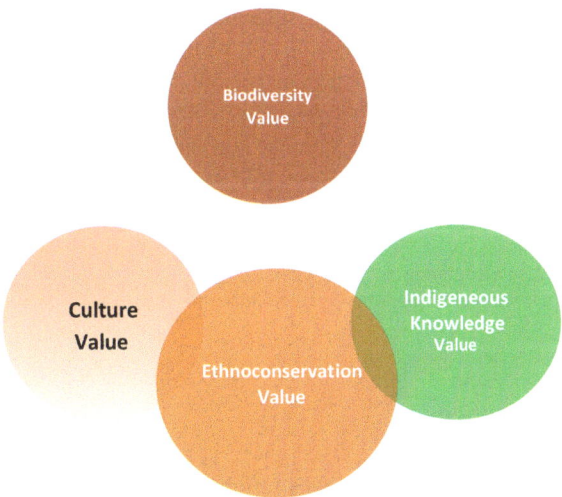

practiced from generation to generation using customary rules, rather than carried out sustainably through the process of family and community education (Tamalene and Almudhar 2017). Figure 10.12 shows the three important values in ethnoconservation concept: biodiversity values, cultural values, and indigenous knowledge values.

The six benefits of ethnoconservation are: (1) Maintaining food stability and wildlife activities in rural areas of certain ethnic groups, (2) Maintaining good relations with the natural and social environment, (3) Minimizing large-scale damage to biodiversity, (4) Creating local laws obeyed by community-related to natural resource conservation, and (5) educating the community, especially the younger generation, to keep protecting the surroundings and serving as the model of hidden curriculum.

In conclusion, this study shows endemic birds on the Halmahera Island have ecological and cultural importance shaped by the Ngora Ma Beno concept. Incorporating this indigenous knowledge creates an approach to bird conservation in which each species' sounds/chants are culturally significant and, therefore, have implications for the birds' ecology. Local people have an important role in protecting biodiversity, especially birds believed to be cultural symbols. Thus, a conservation-based indigenous knowledge approach plays an important role in educating local people, creating a sense of belonging, and enhancing local people's pride in changing behavior and protecting birds. The factors helping species to become recognized and appreciated are not only ecological but also social, personal, cultural, emotional, and economic (Aiyadurai and Banerjee 2020). In addition, tourism increases capacity building in remote communities, which often coincides with opportunities for residents to be trained as guides by international bird conservation organizations in order to promote sustainable tourism (Biggs et al. 2011). Therefore, the integration of ecotourism and cultural tourism is crucial (Harbor and Hunt 2021). Finally, these findings also show that bird ecotourism increases women's role in sustainable bird conservation, as women are typically the managers of ecotourism in the village. Therefore, local women need to be more involved in ecotourism management. This research recommends formation of pro-conservation women's groups in villages, by governments in various countries, as an important part of minimizing bird poaching. Village-based bird conservation practices that maximize the role of women are bound to be a special attraction for ecotourism benefits.

Acknowledgements The authors are grateful to the informants and village heads in the study areas for granting permission to conduct this research.

References

Aiyadurai A, Banerjee S (2020) Bird conservation from obscurity to popularity: a case study of two bird species from Northeast India. Geo J 85(4):901–912. https://doi.org/10.1007/s10708-019-09999-9

Balasubramaniam KN, Marty PR, Arlet ME, Beisner BA, Kaburu SSK, Bliss-Moreau E, Kodandaramaiah U, McCowan B (2020) Impact of anthropogenic factors on affiliative behaviors among bonnet macaques. Am J Phys Anthropol 171(4):704–717. https://doi.org/10.1002/ajpa.24013

Biggs D, Turpie J, Fabricius C, Spenceley A (2011) Africa Linked references are available on JSTOR for this article: The Value of Avitourism for Conservation and Job Creation - An Analysis from South Africa. 9(1):80–90. https://doi.org/10.4103/0972-4923.79198

Chiwanga FE, Mkiramweni NP (2019) Ethno-ornithology and onomastics in the Natta community, Serengeti district, Tanzania. Heliyon 5(10):e02525. https://doi.org/10.1016/j.heliyon.2019.e02525

D'Amico ZTM (2017) Ecotourism's promise and peril. In: Bessa DTBGSMS (ed) Ecotourism's promise and peril. Springer, Cham, pp 97–115. https://doi.org/10.1007/978-3-319-58331-0

Hakim L (2017) Managing biodiversity for a competitive ecotourism industry in tropical developing countries: New opportunities in biological fields. AIP Confer Proceed:1908. https://doi.org/10.1063/1.5012708

Harbor LC, Hunt CA (2021) Indigenous tourism and cultural justice in a Tz'utujil Maya community, Guatemala. J Sustain Tour 29(2–3):214–233. https://doi.org/10.1080/09669582.2020.1770771

Idris DSRPH, Gadong MS, Morni MIR, Wahab AMFSP (2019) Sumbiling eco village: promoting ecotourism in the Temburong district. Green Behav Corporate Social Responsibil Asia:57–63. https://doi.org/10.1108/978-1-78756-683-520191007

Kibria ASMG, Behie A, Costanza R, Groves C, Farrell T (2020) Potentials of community-based ecotourism to improve human wellbeing in Cambodia: an application of millennium ecosystem assessment framework. Inter J Sustain Develop World Ecol 00(00):1–12. https://doi.org/10.1080/13504509.2020.1855606

Maccarthy M, Mary S (2020) Tourism and indigenous peoples setting the stage for indigenous tourism. July, 1–28.

Maldonado JH, del Moreno-Sánchez R, Espinoza S, Bruner A, Garzón N, Myers J (2018) Peace is much more than doves: The economic benefits of bird-based tourism as a result of the peace treaty in Colombia. World Develop 106:78–86. https://doi.org/10.1016/j.worlddev.2018.01.015

Mann S (2016) The Research interview. Res Interv. https://doi.org/10.1057/9781137353368

Markwell K (2018) An assessment of wildlife tourism prospects in Papua New Guinea. Tour Recreat Res 43(2):250–263. https://doi.org/10.1080/02508281.2017.1420008

Nuckel ST (2019) Ecotourism development in indigenous communities: a mapuche case study

Park G, Hotel L, Linhartova V, Republic C, Proceed- I (2019) New trends and issues proceedings on humanities. 6(8):21–29

Steven R, Jones D (2014) Encyclopedia of tourism. Encyclop Tour:1–2. https://doi.org/10.1007/978-3-319-01669-6

Stronza AL, Hunt CA, Fitzgerald LA (2019) Ecotourism for conservation? Ann Rev Environ Res 44:229–253. https://doi.org/10.1146/annurev-environ-101718-033046

Su K, Ren J, Qin Y, Hou Y, Wen Y (2020) Efforts of indigenous knowledge in forest and wildlife conservation: a case study on bulang people in mangba village in yunnan province, china. Forests 11(11):1–16. https://doi.org/10.3390/f11111178

Tamalene MN, Almudhar MHI (2017) Local knowledge of management system of forest eco-system by Togutil Ethnic group on Halmahera Island, Indonesia: Traditional utilization and conservation. Inter J Conservat Sci 8(3):509–518

Tran L, Walter P (2014) Ecotourism, gender and development in northern Vietnam. Annals Tour Res 44(1):116–130. https://doi.org/10.1016/j.annals.2013.09.005

Walter P (2020) Community-based ecotourism projects as living museums. J Ecotour 19(3):233–247. https://doi.org/10.1080/14724049.2019.1689246

Wolf ID, Croft DB, Green RJ (2019) Nature conservation and nature-based tourism: a paradox? Environments - MDPI 6(9):1–22. https://doi.org/10.3390/environments6090104

Zuniga RB (2019) Developing community-based ecotourism in Minalungao national park. Af J Hospital Tour Leisure 8(SpecialEdition):1–10

Index

The manufacturer's authorised representative in the EU is Springer
Nature Customer Service Centre GmbH, Europaplatz 3, 69115 Heidelberg,
Germany. If you have any concerns regarding our products, please
contact ProductSafety@springernature.com

Printed and bound by CPI Group (UK) Ltd, Croydon, CR0 4YY

24/04/2026

02096316-0002